Drones and Support for the Use of Force

James Igoe Walsh and
Marcus Schulzke

University of Michigan Press
Ann Arbor

Published in the United States of America by the
University of Michigan Press
Manufactured in the United States of America
♾ Printed on acid-free paper
First published November 2018

A CIP catalog record for this book is available from the British Library.

Library of Congress Cataloging-in-Publication Data

Names: Walsh, James Igoe, author. | Schulzke, Marcus, author.
Title: Drones and support for the use of force / James Igoe Walsh and Marcus Schulzke.
Description: Ann Arbor : University of Michigan Press, [2018] | Includes bibliographical references and index. |
Identifiers: LCCN 2018023715 (print) | LCCN 2018028500 (ebook) | ISBN 9780472124299 (E-book) | ISBN 9780472131013 (hardcover : alk. paper)
Subjects: LCSH: Drone aircraft—Government policy—United States. | United States—Military policy—Public opinion. | War—Public opinion. | War—Moral and ethical aspects. | Drone aircraft—Moral and ethical aspects. | War and society—United States. | United States—Public opinion.
Classification: LCC UG1242.D7 (ebook) | LCC UG1242.D7 W34 2018 (print) | DDC 358.4/14—dc23
LC record available at https://lccn.loc.gov/2018023715

Cover illustration courtesy Jules2000/Shutterstock.com

Acknowledgments

For comments and suggestions on earlier versions of the manuscript, we thank Max Abrahms, Mary Layton Atkinson, Mia Bloom, Justin Conrad, Graeme Davies, Erik Gartzke, Aaron Hoffman, Michael Horowitz, Jenna Jordan, Sarah Kreps, Cherie Maestas, James A. Piazza, Todd Sandler, John Szmer, and Joseph Young. We are particularly grateful to Victor Asal, who introduced us and encouraged us to collaborate on this book, as well as two anonymous reviewers whose suggestions led to a much-improved final product. We presented parts of the work at the annual meetings of the American Political Science Association, the Military Operations and Law Conference, Tufts University, the University of St. Gallen, the Middlebury Institute for International Studies, and the Pennsylvania State University, and we thank participants for their input. At the University of Michigan Press, Danielle Coty, Elizabeth Demers, and Scott Ham provided important suggestions and guided the manuscript to publication.

Financial support was provided by the University of York and the University of North Carolina at Charlotte. Samantha Luks and Marissa Shih at YouGov ably implemented the experiments, which were approved by the Institutional Review Board at the University of North Carolina at Charlotte. Replication code and data can be found at http://www.jamesigoewalsh.com

Contents

CHAPTER I

Introduction

Choosing War

Political leaders face two key challenges when they decide to use military force: winning the war itself, and winning support at home. In the past two decades, the United States has pursued a technological solution to these problems by developing combat drones—weapons that can both selectively target opponents and minimize the costs and risks of combat. In this book, we seek to understand how this solution to the challenge of military victory also addresses the need for public support for engaging in conflict.

Combat drones have been employed by the United States against insurgent and terrorist groups. These militant organizations are materially weaker than the governments they fight. Their relative weakness leads them to avoid direct military confrontations and to use the civilian population to mask their identities, attempting to pass as noncombatants by eschewing uniforms and by residing in populated areas. A key challenge that authorities face in countering insurgencies is solving this "identification" problem of distinguishing bona fide militants from civilians.[1] Doing so allows the authorities to bring to bear their superior military power. But failure to correctly identify militants means that strikes risk military casualties, mission failure, and civilian harm.[2] This is not only unfortunate but counterproductive, as militants can use their opponent's failure and the deaths of noncombatants to persuade the population that the authorities are indiscriminate and unjust, while the militants can offer protection today and the promise to replace the government in the future.[3]

The United States faced just this problem in its attempts to counter insurgencies in Vietnam, Iraq, and Afghanistan. In Vietnam, areas subject to aerial bombing experienced many civilian casualties and saw declines in government control and increases in influence by militants. In Iraq, inadvertent killings of civilians by the United States led to more militant violence in regions where government influence and support was the weakest. In Afghanistan, popular support for the international military coalition led by the United States declined in villages where its use of force resulted in civilian harm.[4]

Leaders in democracies also need to maintain domestic political support for the conflict by convincing the public that the benefits and likelihood of eventual victory will exceed the human and financial costs. Their ability to do so depends in part on success on the battlefield, but it is difficult to demonstrate with much certainty that counterinsurgency campaigns are "working." Militant movements present few opportunities to decisively defeat the enemy in set-piece battles, and the militants' ability to hide among the population makes it difficult to assess their military capabilities. Even when militants lose major engagements, as they did during the Tet Offensive in Vietnam and the Battles of Fallujah in Iraq, the outcome rarely feels like a victory because the confrontations provide evidence of militants' strength and foster doubts about whether long-term political objectives can be met. Citizens can clearly see the costs of using force against militants—government spending and military casualties, for example—but have difficulty seeing any successes. Mounting financial and human costs, especially military casualties, directly reduce support for countering insurgencies. Vietnam, Iraq, and Afghanistan illustrate this dynamic; in each case, public support for the use of force declined as casualties increased.[5]

Technological developments now allow the creation of weapons, the most prominent of which are armed unmanned aerial vehicles, or "drones," that promise to make it easier to address both challenges. These combat drones have two relevant characteristics when it comes to managing the costs of counterinsurgency and improving the chances of success. The first is selectivity, understood as the ability to identify and strike a well-defined target—such as a particular building, vehicle, or individual—while minimizing harm to nearby noncombatants and civilian infrastructure. Weapons are more selective when they can be actively guided to their targets, such as missiles directed to specific geographic coordinates. Selectivity is also enhanced when the weapon is integrated with multiple streams of intelligence—collected by the weapons platform itself or by other systems, such as satellites or communications intercepts—that provide information about the identity and location of a target. Greater selectivity means that a weapon is better able to solve the identification problem, allowing more effective detection and targeting of hostile militants while timing strikes to reduce the possibility of civilian casualties. Drones hold advantages in terms of both aspects of selectivity, as they are tools for conducting protracted reconnaissance missions to track suspected enemy fighters and for launching attacks using relatively low-yield munitions that reduce the risk of inflicting civilian casualties.

The second characteristic is pilot invulnerability: the capacity to achieve battlefield victories while minimizing the risk that military personnel face when engaging enemy forces. Pilot invulnerability is a function of the range with which a weapon can strike from its operator. In general, the longer the range, the less vulnerable the weapon's operator is to enemy fire. Field artillery, for example, typically has a much longer range than mortars, meaning that soldiers manning the former are, all other things being equal, at less risk than those operating the latter. Similarly, the crews of aircraft armed with guided missiles are less vulnerable than their counterparts flying aircraft that drop simple "gravity" bombs, which have to fly closer to their targets to increase their accuracy.

This invulnerability is a characteristic of the pilot, not of the machine. Drones remove their human operators from danger, but the machines themselves may be attacked and destroyed. Current generation drones are generally more susceptible to attack than aircraft with onboard crews because they tend to fly slower, have limited defensive capabilities, and perform poorly in air-to-air combat against enemy aircraft.[6] Drones may also be hacked, spoofed, or suffer from technical faults that interfere with their control systems. These limitations put the machines at risk of attack or failure, yet the pilots remain invulnerable because they are at such extreme distances from the battlefield that their fate is completely divorced from that of the aircraft. The entire American drone force could be shot down without causing any loss of human life; the same cannot be said of aircraft without onboard crews.

Both characteristics have powerful political consequences, especially when they are combined in the same weapon system. This has not received sufficient attention, in part because until recently the designers of weapons have faced a trade-off between selectivity and pilot invulnerability. To continue the examples introduced above, mortars are typically more selective than artillery but place troops in greater danger of counterattack. Strike aircraft armed with laser-guided bombs can regularly hit specific targets, but this selectivity declines when they fly at higher altitudes to avoid anti-aircraft fire. For this reason, much analysis has focused on either selectivity or pilot invulnerability, while assuming that these goals cannot be realized simultaneously.[7] There is a robust debate dating back to the 1920s, for example, about how selectivity influences political and military outcomes.[8] Other works have considered how pilot invulnerability reduces the costs and risks of using force, for example by allowing attacks with long-range aircraft compared to the use of ground troops.[9]

Recent technological developments have sharply narrowed the trade-off between selectivity and pilot invulnerability. Combat drones—remotely piloted aircraft armed with air-to-ground missiles—are the best example of a weapon system that incorporates both characteristics. Drones are more selective and offer their operators greater safety than their most similar weapon system, the strike aircraft with an onboard crew. Both are armed with accurate, guided munitions and can collect intelligence about potential targets from onboard sensors. Drones have the advantage of being able to loiter for much longer periods, allowing more time for the integration of intelligence, the positive identification of targets, and the selection of a time to strike that will maximize damage to the target while minimizing risks to noncombatants. Modern strike aircraft have considerable ability to avoid taking enemy fire, including high-speed, defensive weapons, and in some cases a "stealthy" design. But their crews are always at some risk of being shot down or experiencing a malfunction, while the pilots of a drone may be located thousands of miles from the battlefield and immune from physical harm. Importantly, drones represent an improvement in *both* selectivity and pilot invulnerability over other weapons systems. Piloted aircraft, for example, can increase the selectivity of their attacks by flying closer to their targets, but this increases the vulnerability of their crews to ground fire or accidents. Drones require no such trade-off; they allow attacks that are more selective while eliminating risks of physical harm to military personnel.

The ability of more selective weapons, such as drones, to mitigate the political difficulties of war—winning on the battlefield and securing domestic support—has been debated since the advent of air power in the early twentieth century.[10] Campaigns that shaped this debate include the British use of air power to counter rebels in northern Iraq and in Waziristan in the 1920s and the United States' extensive bombing against insurgent Viet Cong units during the Vietnam War.[11] Most of these attacks were carried out with gravity bombs dropped from manned aircraft, which could not be directed to their targets with much precision and which placed their crews at risk. Later assessments concluded that they had little effect on, or actually strengthened, the insurgents they targeted.[12] The use of truly selective weapons has been much less frequent or sustained. The United States launched cruise missiles against al Qaeda bases in Sudan and Afghanistan after attacks on American embassies in 1998, but these attacks were not maintained for long enough to undermine the group. Israel has used missiles fired from helicopters, fixed-wing aircraft, and drones, as well as snipers and ground troops, in a campaign of "targeted killings" directed against armed Palestinian groups in the West Bank and Gaza Strip, and the United States used similar technologies of violence in Iraq and Afghanistan. These campaigns were certainly more selective than their predecessors, although their military effectiveness has been debated, and they continued to place military personnel at some risk of harm.[13]

Drones may fundamentally alter these trade-offs and more generally the costs and benefits of using armed force. Scholars of international politics have long thought that the costs of conflict influence potential combatants' willingness to take up arms.[14] Recent work uses as its starting point the assumption that states and other actors who have different interests and preferences bargain with each other to resolve these differences. Using force, or threatening to do so, is one strategy an actor can adopt to press its opponent to make concessions. But even successful military action is costly; it risks the lives of military personnel and civilians, is expensive in financial terms, and reduces the capacity to use force in other theaters or against other opponents. Before resorting to military force, states and other actors subtract their estimates of these costs from the benefits they expect to accrue from successful coercion of a foe. This creates a "bargaining range," understood as negotiated agreements that both sides prefer to incurring the costs and risks of war. When the bargaining range is larger, there are more potential bargains that the sides might prefer to war.[15]

The costs of war are an important influence on the size of the bargaining range; as these costs change, so does the range of peaceful settlements that both sides would prefer to armed conflict. Innovation in military technology and doctrine is one way that the costs of war change. Innovations that make war more costly should, ceteris paribus, create a larger bargaining range and thus reduce the likelihood of war. A good example of this is the effect of nuclear weapons on interstate conflict; many conclude that, under the right conditions, nuclear weapons can prevent conflict by hugely raising its costs.[16]

Nuclear weapons increase the costs of war, while drones reduce them. This means that possession of armed drones could strengthen incentives for an actor to start or to sustain conflicts. The pilot invulnerability that drones allow eliminates a key cost of military action: the military casualties that undermine domestic political support for the conflict. Their selectivity means that weapons launched from drones are more likely to achieve their battlefield objectives, increasing the costs imposed on the opposing side. Selectivity also lowers the chances that releasing such weapons will result in harm to civilians, which, like military casualties, can make political actors at home and abroad less willing to support the conflict. Drone operations further reduce the need for large forward military bases, lowering (but not eliminating) the financial costs of military operations and the need to rely on allies, who may demand concessions in exchange for their cooperation. These reductions in the cost of war are particularly strong for combatants that possess drones and that face foes unable to destroy these weapons in combat or to easily impose other costs on the adversary. This has been the situation for many of the conflicts involving terrorist and militant organizations against whom the United States has used armed drones since 2002.[17]

Drones have substantially reduced the costs of conflict for the United States. The concern is that this has made armed conflict a more attractive option for the United States. In contrast with nuclear weapons, which may have reduced the likelihood of interstate war, the development of drone technology may create incentives for more and longer conflicts. The use of combat drones by the United States in the twenty-first century, then, marks an important change from these earlier conflicts. The United States and other countries have been developing precision-guided munitions in earnest since the 1970s. But until recently, these weapons were not used regularly enough to permit a sustained analysis of their effects. For this reason, most assessments of selective weapons have been anecdotal. The drone campaigns of the past decade and a half have involved hundreds of strikes against militant organizations, most in the Federally Administered Tribal

Areas of Pakistan, with additional strikes in Yemen, Somalia, Iraq, and Syria.[18] This large number of attacks over a long period of time provides the first opportunity to assess in a systematic way how the sustained use of such weapons influences public support for the use of force.

Support for the Use of Force

This book seeks to understand how the development of combat drones influences popular support for the use of force by the United States. Investigating the relationship between public opinion and foreign policy has a long history in political science.[19] Much of the recent work in this area can be grouped into three broad schools of thought. A first school, and the one with which the book engages most directly, focuses on how individuals assess the costs and benefits of using military force or engaging in other foreign policy actions; public support rises as the benefits increase, the costs decrease, or both. Benefits of using force can include deterring states (or nonstate actors, such as terrorist or militant groups) that threaten the security of the United States, coercing states and nonstate actors into taking actions that benefit the United States but that those actors would otherwise prefer to avoid, preventing such actors from committing mass killings or other atrocities, or supporting allied states against the threats they face. Considerable research shows that the public attaches different weights to the value of each of these "principal policy objectives," and that these weights can change as the threats to American interests and values vary over time. For example, deterring aggression by powerful states was especially important during the Cold War, when this was the principal goal of American foreign policy, while the importance of countering terrorism has increased since the terrorist attacks of September 11, 2001.[20]

According to this line of reasoning, the public balances these potential benefits against the costs of using force. The cost that has received the most attention is military casualties, which have been shown to be a consistent and important influence on such attitudes. A large body of research has concluded that the occurrence, rate, home state, race and ethnicity, socioeconomic status, timing, and framing of military casualties influence Americans' attitudes.[21] Another important cost is the possibility that military action will fail to secure American interests. Influential recent work concludes that the American public is willing to tolerate the risk of military casualties if doing so ensures that important American interests are secured.[22] Less work has investigated how the public assesses the cost of civilian casualties that result from American military action. Some conclude that such casualties play little role in the citizens' cost-benefit calculations, although more recent work suggests that many individuals attach considerable weight to avoiding harm to noncombatants.[23]

This book seeks to understand how the use of combat drones influences these cost-benefit calculations. This is an important contribution to the study of public attitudes. Most of the research to date on public opinion and the use of force assumes, often implicitly, that military action will place American service members at risk of harm on the battlefield. Perhaps the most widely discussed consequence of combat drones is that pilot invulnerability eliminates military casualties. Drones' selectivity also has the potential to reduce many of the costs of combat. If drones can target the leaders of insurgent and terrorist organizations, their use could undermine these combatants' political and military effectiveness, leading to shorter or less bloody conflicts. The ability of drones to reduce civilian casualties compared to other, less selective forms of military force could make the deaths of noncombatants less consequential for the American public. If drones reduce these costs of combat, then this would have important implications for decisions to initiate and to continue conflicts. If the United States is able to pay much lower costs for engaging in combat, this could increase the incentive to resort to arms and decrease the willingness to find peaceful solutions. This in turn may lead the country to engage in more, and possibly more reckless, wars to achieve its aims. Our goal is to assess the degree to which drones lead individuals to reassess these costs of conflict and how such reassessments influence their support for military action.

A second school emphasizes that individuals' preferences and opinions are heavily influenced by the information provided to them by the media and by political elites. According to the most influential version of this approach, when elites across the political spectrum agree on the wisdom of using military force, citizens who follow politics align their views with those of the country's political leaders. But when these leaders hold conflicting preferences about foreign policy, individuals express support for the policies espoused by leaders who share their partisan identity. Related work has explored how the preferences of other elite actors, such as international organizations and nongovernmental organizations, as well as the framing of conflicts by the media, also drive individuals' perceptions and preferences.[24]

Although we focus on public opinion formation in direct response to prospective attacks, rather than on the role of elite influence, our results indirectly address this perspective. Reducing the perceived costs of fighting for the public may reduce the perceived costs for elites as well. That is to say, elites may themselves subscribe to the narrative of drones being effective low-cost weapons that can be used more readily than other types of military force. If elites adopt this viewpoint, they may be more credible and influential sources of guidance for the American public. Confidence in drones will also increase the likelihood that they will be portrayed as a viable attack strategy when the country is contemplating war. While the findings presented in subsequent chapters arise from a different starting position than much of the elite-focused work on public opinion, we demonstrate at multiple points that the interplay between elite and citizen preferences could also influence the domestic politics of employing combat drones. For example, in chapter 4 on principal policy objectives, we discuss how political leaders have powerful incentives to frame conflicts in terms of objectives that receive more public support. Our findings also have implications for what strategies elites may adopt to frame information effectively.

A third school of thought begins with the premise that individuals have underlying values about the utility and morality of military force and the importance of international collaboration as a means for achieving national objectives. This school of thought holds that general foreign policy attitudes play an important role in shaping individuals' understandings of and preferences regarding military action in a wide range of specific circumstances.[25] We have little doubt that such attitudes could influence how Americans think about combat drones. It seems plausible to expect that, for example, individuals whose attitudes about foreign policy are characterized by "militant assertiveness" would be more willing to countenance drone strikes. Our focus is on the more proximate factors surrounding combat drones, such as their selectivity and pilot invulnerability, as these characteristics are the most novel aspects of the technology and exercise a direct influence over the costs and benefits of decisions to use military force. However, our experiments indirectly address questions about the influence of existing values. In particular, we find that Democrats assign less utility to using force than do Republicans. This likely reflects underlying differences in partisans' assessments of the utility and morality of military force. We find little evidence of differences in partisan preferences for different types of military force; both Democrats and Republicans prefer drone strikes to the use of ground troops. This suggests that while overarching political ideologies matter for assessments of the value of military action, there is considerable consensus preferring attacks from more selective and invulnerable platforms that reduce the costs of conflict for the United States.

History of Drone Warfare

Drones respond to the challenges of managing the costs of war and more precisely targeting enemies. These challenges have become more prominent in recent counterinsurgencies, but they are not entirely new. Remote weapons designed to increase pilot invulnerability and selectivity predate the current generation of drones by several decades. Tracing the evolution of these weapons from the Second World War to the War on Terror is important for four reasons. First, it offers insight into the links between the goals of improving pilot invulnerability and selectivity. Second, it shows how drones have been used to not only manage the risks of fighting, but to also avoid displeasing an American public often seen as being reluctant to wage costly wars, especially in the post-Vietnam era. Third, it shows the extent of recent drone operations throughout the War on Terror and the ongoing race to produce new drone variants, which in turn suggests that drones will continue to play a central role in future American military operations. Finally, this history clarifies the moral and legal questions that drones pose, especially those relating to unintended consequences of lowering inhibitions against fighting, which we test throughout the book.

Early drone variants were produced by both sides during the Second World War. Germany experimented with unmanned ground vehicles (UGVs) by introducing the Goliath remote-controlled mine, which was designed to be driven under enemy tanks before exploding. The Allies tested their own drones, though they were more interested in unmanned aerial vehicles (UAVs) and initially only saw its Radioplane aircraft as training tools to give pilots practice with air-to-air combat.[26] These and other early applications of remote weaponry pointed the way toward future developments in unmanned warfare, yet they showed that the technology had not advanced to a stage that would make drones feasible weapons. Remote mines were single-use devices that had to be operated at short range, thereby failing to provide operator invulnerability. They were also unreliable and difficult to control, which made them a poor addition to Germany's arsenal. Target UAVs were more successful, becoming a fixture of training for United States Air Force pilots during the Cold War and gradually transitioning into use as reconnaissance aircraft as their control and communication systems became more advanced.

Among the earliest drones in the sense of being able to replace a manned vehicle in precise and low-risk attacks was the Ryan Firebee. Originally designed as a target for pilots in training, it was later adapted for ISR (intelligence, surveillance, and reconnaissance) in the form of the Fire Fly and Lightning Bug variants. From 1964 to 1974, Firebees flew 3,435 sorties over Southeast Asia.[27] Some were shot down over North Vietnam, yet they had a survival rate averaging around 84 percent, reaching as high as 90 percent near the end of the war.[28] These aircraft were designed to be recoverable and even carried countermeasures to interfere with enemy targeting systems, which allowed them to take over at a low cost some of the missions that would have endangered human pilots. Still, they were primitive compared to modern drones, and technical limitations prevented them from being used on a larger scale. Among the most serious shortcomings were that they could not take off and land on their own, depending on assistance from manned aircraft for launch and recovery, and they had to be controlled by nearby DC-130 director planes.[29] Thus, the reconnaissance drones failed to provide pilot invulnerability—only reducing the risk to pilots by distancing them from their drones while still forcing them to remain in the battlespace—and lacked the capacity to conduct selective attacks.

The MQ-1 Predator was first deployed in 1995 to provide surveillance in the Balkans. The small, light aircraft were ideal for unarmed reconnaissance and were used in that capacity over the following decade. Their utility compared to manned aircraft became clear when Scott O'Grady was shot down over Bosnia on July 2, 1995. O'Grady spent six days evading enemy fighters, with his struggle to avoid capture and his rescue becoming a major media event. He survived the ordeal, yet it provoked fears that ostensibly low-cost humanitarian interventions conducted with the utmost care could still result in American military casualties. The incident built on fears that public support for fighting could drop sharply with the loss of even a single American life. On August 11, 1995, a Predator drone was shot down over Bosnia, but this incident attracted much less attention and lacked the human element that had made O'Grady such a compelling figure. The apparently limited concern over the crash seemed to confirm that pilot invulnerability could lower the costs of war and help to avert potentially disastrous drops in public support. Predators returned to service in 1999 during the intervention in Kosovo. Once again, they were used to locate enemy positions concealed in forests or towns, flying at low altitudes where manned aircraft would not have been safe from ground fire.[30] The importance of pilot invulnerability was evident, yet UAVs were still only seen as facilitating selective attacks and not as weapons platforms in their own right.

Predator drones began operating in Afghanistan in 2000 as part of the Central Intelligence Agency's (CIA) search for Osama bin Laden following al Qaeda attacks on US embassies in Kenya and Tanzania in 1998. The plan was to use a drone to direct a cruise missile to the target once he was discovered. A man matching bin Laden's description was found, but attack orders were not issued. The mission faced a serious challenge when it came to ensuring that the target would remain in the same location while a missile traveled hundreds of miles to reach him. With this difficulty in mind, the CIA sought permission to arm a Predator with Hellfire missiles.[31] Advocates claimed that mounting weapons directly on UAVs was the best way to eliminate a delay between missile launch and impact, which could otherwise allow a target to escape or endanger innocent people who might wander into the attack site. Thus, the impetus for arming Predator drones came from a desire to combine the pilot invulnerability of surveillance aircraft with weapons that were more selective than cruise missiles.

The first targeted killing with a Predator UAV took place on October 7, 2001, and was directed against Mullah Omar, the leader of the Taliban. The plan was to kill several bodyguards, which would then draw Mullah Omar into the open where a second missile would be able to strike him. However, the target was able to escape, leaving only a few low-ranking bodyguards as casualties. Control of the Predator used in the attack was contested from the start, as "the drone was remotely piloted by USAF operators, working from a mobile station in the carpark of the CIA's headquarters in Langley."[32] The CIA had overall authority over the strike, but the Air Force insisted on being involved, and the CIA was obliged to assent because of collaborative arrangements between the military and intelligence services that were designed to facilitate the War on Terror. Failure to kill the target escalated the tensions between the services involved and fueled a protracted struggle over the management of lethal UAV operations.

Over the following months the CIA began arming its Predators and using them to conduct additional strikes against suspected al Qaeda and Taliban leaders in Afghanistan. With only around 40 missiles fired in October and November 2001 compared to more than 6,500 strikes by manned aircraft, the Predator's contribution to the war effort was modest.[33] Singer estimates that Predators had only launched independent attacks on 115 targets in Afghanistan by the end of their first year in the conflict.[34] They were more commonly used to mark targets for manned aircraft, assisting in around 525 strikes during the first two months of fighting in Afghanistan.[35] The Bureau of Investigative Journalism estimates that just 52 strikes were carried out during President Bush's time in office, resulting in 416 deaths, including approximately 167 civilians.[36] Reliance on armed drones therefore developed slowly throughout the Bush administration's time managing the program.

In terms of numbers alone, the scale of drone strikes was comparatively modest from 2001 to 2008, but the geographical scope expanded considerably. Targeted killing operations spread to Yemen in 2002, when a suspected participant in the attack on the USS *Cole*, Qa'id Salim Sinan al-Harithi, was assassinated.[37] He and five other members of al Qaeda were killed as they were driving, which further demonstrated the drones precision by showing that they could hit moving vehicles. Operations in Pakistan began in June 2004 with an attack in South Waziristan. This killed Nek Mohammed, a senior Taliban leader, but also killed two young boys and several other bystanders who may have been civilians. The attack highlighted how opaque these operations were, particularly when it came to attribution, as Pakistan's government initially claimed responsibility for the attack. It was also evidence of the tenuous relationship between Pakistan and the United States. The Pakistani government had authorized the attack following its failure to capture Waziristan with ground forces, and yet it still publicly criticized American actions. Attacks like these fueled concerns over the morality of drone warfare, and especially of the unintended consequences of using weapons that could improve precision while also making pilots invulnerable to attack.

The Bush administration's drone strikes increased sharply in 2008. It launched around 36 attacks in Pakistan, where the previous annual high had only been four. This trend continued under the Obama administration, with 54 strikes in 2009, more than in all previous years combined. Drone operations in Pakistan reached their peak in 2010, with roughly 122 attacks. The height of activities in Yemen came two years later with around 41 strikes.[38] The Bureau of Investigative Journalism estimates that around 421 UAV attacks have occurred in Pakistan since the start of operations in 2004, resulting in 2,476–3,989 people killed and 1,158–1,738 injured.[39] Since 2002, between 107 and 127 strikes have taken place in Yemen, with 492–725 people killed and 94–223 wounded. The Bureau of Investigative Journalism also estimates that the civilian death rates among the total casualties in drone strikes are between 11 percent and 39 percent in Pakistan and between 9 percent and 21 percent in Yemen.[40]

Prior to mid-2008, the United States launched drone strikes in Pakistan when it was able to identify the location of a named individual leader of a militant group. These "personality strikes" were few in number; only two were launched in 2006 and four in 2007. These rules of engagement were altered in early 2008 to allow attacks against groups of armed men that bore the "signatures" of militants, as long as no civilians were nearby. Off-the-record explanations by American and Pakistani officials were summarized in the following terms:

> Instead of having to confirm the identity of a suspected militant leader before attacking, this shift allowed American operators to strike convoys of vehicles that bear the characteristics of al-Qaeda or Taliban leaders on the run, for instance, so long as the risk of civilian casualties is judged to be low.[41]

Another journalist used the following examples to describe the new procedures: attacks from drones could be launched "based solely on intelligence indicating patterns of suspicious behavior, such as imagery showing militants gathering at known al-Qaeda compounds or unloading explosives."[42] The emergence of signature strikes showed growing confidence drones as the key to killing terrorist leaders with minimal American losses, but it contributed to moral concerns that drones were making war too easy.

Shortly after this shift in policy, the number of drone strikes increased to 34 in 2008, 53 in 2009, and 119 in 2010. This reliance on patterns of behavior that resemble those of militants the United States wishes to target, but who cannot be positively identified, soon attracted concerns from within the United States government. One American official critical of this policy stated that "it bothers me when they say that there were seven guys, so they must all be militants. They count the corpses and they're not really sure who they are."[43] Internal, classified assessments of most drone strikes conducted from 2006 to 2008 and in 2010 and 2011 call many of the unidentified individuals "foreign fighters" or "other insurgents" or members of militant groups, such as the Pakistan Taliban and Lashkar i Jhangvi, who have not regularly targeted the United States.[44]

Beginning in early 2012, the United States developed new procedures to govern drone strikes in Pakistan. Many of these focused on ensuring that no civilians were likely to be harmed by drone strikes, as it became clear that civilian deaths were attracting increasing criticism from political parties and other groups within Pakistan. That is to say, the policies were driven by a desire to further improve the selectivity of drones with a policy-oriented approach to how they were employed. This move to tighten and institutionalize the conditions under which it was permissible to launch drone strikes was associated with a decline in their use to 73 attacks in 2011 and 48 in 2012. The shift also suggests that the rules governing the use of drones in earlier years was considerably more elastic and that more attacks were justified by the activities or location of the target than by precise intelligence on the target's affiliation.[45]

Although it has attracted much less media coverage than other fronts in the War on Terror, the Philippines have been the site of at least one attack. The first reported incident came in 2006, with a failed attempt to kill Umar Patek, one of the terrorists responsible for the Bali bombing in 2002.[46] However, this incident has not been confirmed by any official sources and it remains unclear whether claims of American involvement are accurate. Some say that the February 2012 attack against members of Abu Sayyaf and Jemaah Islamiyah was the first drone strike in the Philippines.[47] Given this degree of uncertainty, it is impossible to say how many strikes have been carried out and how many casualties inflicted.

In Pakistan, Yemen, and Somalia, the United States used drones in conflicts to which it was not an official or declared party. Drone strikes have also taken place alongside other types of American military action in conflicts where the United States is a combatant. The numbers for Afghanistan, Iraq, and Syria are also difficult to discern. With ongoing fighting in these countries, there are challenges in identifying the source of some attacks based on publicly available reports. It is often difficult to tell when a strike was conducted by a drone and when a manned aircraft was involved, especially when reports often refer to both simply as "air strikes." Nevertheless, it is clear that UAV operations in both countries were extensive. One report finds that there were at least 333 attacks in 2012 alone.[48] A report published in 2014 found that "Afghanistan is the most heavily drone-bombed country in the world" with "over 1,000 known to have hit the country in the past 13 years."[49] And these strikes do not come from the United States alone. The British government reported flying over 100,000 hours in Afghanistan as of 2012, firing 349 missiles.[50]

As the Obama administration came to rely more heavily on drones to kill foreign militants, it also spread operations into new areas. The first drone strike in Somalia came in June 2011, wounding several members of al Shabab and foreign militias. Activity in Somalia has consistently been much lower than in other theaters, reaching a peak of around five strikes in the first half of 2016.[51] Between 15 and 19 strikes were carried out in Somalia, resulting in between 25 and 108 deaths and between 2 to 7 injuries. Once again the attack was preceded ongoing unarmed reconnaissance since at least 2009, when an unarmed drone was shot down. This was also reportedly the first time the CIA and Joint Special Operations Command collaborated on a drone strike.[52]

Other operations were conducted in Libya. The Pentagon acknowledged carrying out 145 drone attacks between April 23 and October 20, 2011. Drones from the United States and the United Kingdom continued patrols over Libya during the following years, occasionally launching attacks. The operations were justified as part of the effort to monitor first al Qaeda and then Islamic State operatives. In November 2015 drones killed Abu Nabil, the leader of Islamic State in Libya.[53] And involvement seems set to increase with the United States acquiring an airfield in Sicily from which to conduct drone missions in support of special operations forces working in Libya.

It is useful to reflect on this history with help from research on military innovation. That literature shows that technological and organizational changes can come from a range of different sources, including pressure from civilian politicians, inter-service dynamics, institutional cultures, and bottom-up initiatives from soldiers themselves.[54] The impetus for change and the overall process vary considerably depending on the technology in question, with larger projects tending to require a top-down approach and field modifications or new uses for existing equipment being pioneered through bottom-up approaches. A military conservatism that frequently impedes technological innovation of all types was responsible for limiting enthusiasm for drones throughout the Cold War and reluctance to arm them. Horowitz, Kreps, and Fuhrmann argue that drones offer relatively few advantages during conventional interstate wars because they are more susceptible to enemy aircraft and air defense systems than aircraft with onboard pilots. Drones become far more attractive during asymmetric conflicts in which the enemy lacks anti-aircraft capacities.[55] As long as the Air Force was preoccupied with a conventional war-fighting mission, there was little to gain from arming drones. It took a new mission and external pressure from intelligence agencies and the highest levels of government to overcome this conservatism.

The rise of drones from relatively obscure reconnaissance aircraft to central weapons in America's War on Terror was guided by the novel incentives that policymakers were responding to at the time, especially the challenges associated with hunting suspected terrorists. Intelligence collection and attack functions had to be linked more closely, and it became essential to direct violence narrowly at specific individuals. Once drones were armed, the early innovations were largely driven by inter-service dynamics. The Central Intelligence Agency and the Air Force collaborated on missions while also striving to develop the foremost targeted killing capacities. The Air Force had a cultural bias against launching remote attacks because these conflict with a sense of honor that depends on physical presence on the battlefield.[56] However, competition for resources and missions is a powerful incentive,[57] which compelled it to begin arming its own aircraft even though this conflicted with the institutional culture.[58]

Drones were used sparingly early in the War on Terror because the initial motives for arming drones had limited reach. The central targets were a relatively small number of high-profile terrorists, so the scale of operations was naturally limited. A more concerted effort to employ drones only emerged when members of the Bush and Obama administrations, along with supporters in Congress, decided to rethink the country's strategy. Barry Posen argues that "Even in its own war, a military organization can misperceive the implications of a new technology unless the lessons are very stark."[59] This is borne out by drones, as the military and intelligence agencies had developed a revolutionary new tool as a counterterrorism expedient without this initially having much influence on the US counterterrorism strategy.

Posen contends that major failures cause civilian politicians to intervene in military affairs and force technological change, which explains why the number of strikes rose sharply in 2008 and the scope of targeting parameters increased. By this point, the wars in Iraq and Afghanistan were losing support and the quest to kill Osama bin Laden was failing. Efforts to revive the war effort through troop surges led to protracted partisan disagreements.[60] Drones offered an attractive new approach without further risk to American military personnel at a time when public support was fragile, and they have since held the promise of sustaining counterterrorism operations around the world without the risk of committing large numbers of ground forces. This history of political incentives driving drone innovation highlights the importance of understanding what these incentives are and how they will continue to shape the use of drones in future conflicts.

As this brief history of drone operations illustrates, UAVs are versatile machines, capable of ISR, ground support, targeted killing, and even air-to-air combat against enemy aircraft (though this capability has received relatively little testing). Not all UAV missions are targeted killings. The drones operating in Afghanistan and Iraq also provide support for troops on the ground. Former drone operator Matt Martin describes using a Predator drone to search for and attack mortar crews that were shelling American airbases in Iraq.[61] Drones are also not the only means of conducting targeted killings. Various manned aircraft, cruise missiles, and special operations forces have been involved in these missions, even operating beyond established war zones with their attacks taking place in Pakistan, Yemen, and Somalia. This is evidence that drones are not uniquely connected to any particular mission type and that policymakers and military commanders have a range of attack options available to them. Building on this, one of the central themes of this book will be how members of the civilian public react to policymakers' choices about what weapons and tactics are most appropriate.

Drones come in dozens of variants that perform a broad range of roles, from long-range targeted killing operations to short-range tactical surveillance to bomb disposal. The ScanEagle, RQ-14 Dragon Eye, and RQ-11 Raven each provide short-range unmanned reconnaissance, allowing ground forces to check potential ambush sites or gain greater situational awareness via an overhead view of combat areas. Drones like the PackBot and TALON were employed to defuse or detonate improvised explosive devices (IEDs). Like armed UAVs, these machines owe much of their popularity to their applications in the War on Terror. The PackBot was first used to help in 9/11 recovery efforts.[62] Reliance on it and other ground-based UGVs has been accelerated by the demands of finding ways to protect soldiers in counterinsurgency operations, especially in urban environments.

Since 2001 the US military has used the RQ-4 Global Hawk to conduct surveillance around the world. The aircraft can carry an array of different ISR packages, operates at heights of 60,000 feet, and is able to fly continuously without refueling for more than 25 hours.[63] The RQ-170 Sentinel is a smaller and lighter drone that is likewise used for covert reconnaissance. This machine achieved prominence when one crashed in Iran in 2011, yet it remains shrouded in secrecy and little is known about its capacities. Reconnaissance UAVs face criticisms for infringing on privacy rights, violating state sovereignty, and being dangerously unreliable because of their high crash rate compared to manned aircraft. Nevertheless, the fact that they are not used to kill means that they generally evade media coverage and are marginalized in the ongoing debate over drone warfare, which is typically concerned with lethal machines. Our focus is on UAVs that are used in targeted killing and ground support operations, with the MQ-1 Predator and MQ-9 Reaper foremost among these.

The Predator was designed for reconnaissance, not combat, and was only gradually adapted for fighting. Those responsible for drone operations in the CIA saw arming Predators with AGM-114 Hellfire missiles as a practical necessity—the only way to quickly respond to the targets located when conducting surveillance. Because Predators are extremely light (weighing only around 1,030 lbs. empty and 2,250 lbs. when fully loaded) and were designed for reconnaissance, they are unable to carry the heavier weapons that are typically mounted on manned aircraft and are limited to two Hellfire missiles. The Reaper improves on this with its capacity to carry multiple types of munitions, including GBU-12 Paveway II laser-guided bombs, GBU-38 Joint Direct Attack Munitions (JDAMs), AIM-9 Sidewinder missiles, and the AGM-114 Hellfire II missiles that are used by the Predator. Moreover, Reapers can carry heavier loads, enabling them to operate longer and to attack more targets. These capacities are not always used in practice. Sidewinder missiles are designed for air-to-air combat, in which no Reaper has been involved. Outfitting the aircraft with JDAMs is difficult and was only successfully tested in 2017.[64] This leaves Reapers depending primarily on GBU-12s and AGM-114s, though still with the significant advantage of being able to carry heavier munition loads than its predecessor.

Understanding How Drones Influence Support for the Use of Force

Over time, drone technology has become increasingly selective and has made operators largely invulnerable to enemy fire. Our goal is to assess how these developments influence the public's willingness to support the use of drones on the battlefield. One seemingly straightforward source of answers is public opinion surveys. Surveys asking about support for the use of drones in combat, or about the utility of different types of military force, are conducted frequently. Such surveys, though, are not intended to shed much light on the research questions motivating this book, leading us to rely primarily on survey experiments we designed with this specific goal in mind.

Consider, for example, the data in table 1.1, which summarizes the support for drone strikes among representative samples of Americans during 2012 and 2013—when the drone campaign received significant and often negative coverage in the media—in surveys carried out by different news and survey organizations. In every poll, a majority—and in some case a large majority—express support for drone strikes overseas against targets described as terrorists or militants. Drone strikes, then, appear to have wide support. But this tells us little about *why* drone warfare is popular. We do not know if the public supports drones because they are seen as more accurate or effective weapons, or because they eliminate concerns about military casualties, or because the foes they target are viewed as especially dangerous, or for other reasons. Furthermore, these high levels of support for combat drones are sensitive to how the survey questions were asked.[65]

One issue that public opinion surveys could address, at least in principle, is support for drones compared to other policies, such as relying more heavily on diplomatic efforts or different types of military force, such as ground troops. A window on such issues is provided by the public debate about the wisdom of American action against Islamic State militants during the summer and fall of 2014. This occurred shortly after the Islamic State had seized control of large parts of northern and western Iraq and eastern Syria and had publicly executed an American journalist. Political leaders debated both the wisdom of intervention as well as the form any such intervention should take. Much of this debate centered on the type of military action, if any, the United States should undertake. A number of public opinion organizations polled representative samples of the American public and asked the degree to which they favored or opposed a range of steps being considered by the United States.

Consider the data in the top panel of table 1.2, which summarizes responses to questions about different types of intervention in public opinion surveys conducted during this period. A sizable majority of respondents favored air strikes in Syria and in Iraq, while far fewer supported the introduction of American ground troops. This is consistent with the argument that technologies that reduce the costs of conflict by placing fewer military personnel at risk of harm, such as drones and air power, lead to increased support for the use of force. The lower panel of

TABLE 1.1. Support for Drone Strikes

Date	Organization	Support
February 2012	Washington Post/ABC	83
February 2012	Rasmussen	76
July 2012	Pew Research Center	55
February 2013	Pew Research Center	56
February 2013	CBS	70
February 2013	New York Times/CBS	71
February 2013	NBC/Wall Street Journal	66
February 2013	Fox News	74
March 2013	Gallup	65
March 2013	YouTube/Huffington Post	53
April 2013	New York Times/CBS	70
May 2013	New York Times/CBS	72
May 2013	NBC/Wall Street Journal	64

Note: Support is the percentage of respondents indicating some or strong support for drone strikes against suspected terrorists overseas. For details, see Alyssa Brown and Frank Newport, "In U.S., 65% Support Drone Attacks on Terrorists Abroad," *Gallup*, March 25, 2013; Sarah Dutton et al., "Poll: 45% Approve of Obama's Handling of the Economy—CBS News," *CBS News*, February 12, 2013; Rasmussen Reports, "Voters Are Gung-Ho for Use of Drones But Not Over the United States," February 24, 2012; Megan Thee-Brenan, "Americans Wary on Syria and North Korea Intervention, Poll Finds," *New York Times*, April 30, 2013; Scott Wilson and Jon Cohen, "Poll Finds Broad Support for Obama's Counterterrorism Policies," *Washington Post*, February 08, 2012; Micah Zenko, "U.S. Public Opinion on Drone Strikes," *Power, Politics, and Preventive Action*, 2013, http://blogs.cfr.org/zenko/2013/03/18/u-s-public-opinion-on-drone-strikes/; for a discussion, see Tom McCauley, "US Public Support for Drone Strikes Against Asymmetric Enemies Abroad: Poll Trends in 2013," *Dynamics of Asymmetric Conflict* (n.d.).

table 1.2 depicts responses to questions that asked respondents if they favored the use of drones and manned aircraft in striking the Islamic State. This pair of questions is especially useful for our purposes, since it directly compares the utility that members of the public attach to different types of military action.

Drone strikes received considerably more support than attacks from manned aircraft, which is also consistent with the argument that reducing the costs of war increases support for the use of force. Unfortunately for our purposes, neither of these surveys included questions asking about support for drone strikes, air strikes from piloted aircraft, *and* the use of ground troops, making it impossible to know how the public compares these different forms of military action. Furthermore, the questions in this and most public opinion surveys are intended to measure support for various options, not to assess directly why respondents express the preferences that they do. Yet doing so is important for understanding how changes in the cost of conflict influence public support for the use of force. In the next section, we argue that an experimental approach is better-suited for this purpose, and we summarize results from a series of experiments that are designed to directly test such propositions.

While survey research assembles a group of respondents who are representative of some larger population (such as adults in the United States) and asks them all the same questions, a survey experiment assigns respondents randomly to control and treatment conditions. In our work, these conditions take the form of providing respondents with different information about a planned military attack. For example, in a version of the

TABLE 1.2. Public Attitudes Toward Military Action against the Islamic State

	Favor	Oppose	Don't know/ No answer
Air strikes in Iraq	71	21	8
Air strikes in Syria	69	21	9
Military advisors to Iraq	66	29	6
Train and equip Rebels	48	40	11
US ground troops	39	55	6
Drone strikes	56	38	6
Manned aircraft	43	51	6

Note: Data from CBS/*New York Times* public opinion surveys released September 17, 2014 (*top panel*) and June 23, 2014 (*bottom panel*). Totals do not equal 100 due to rounding.

"attack type" experiment we use in most chapters of the book, respondents are randomly assigned to read online one of three mock news stories describing plans by the United States to use military force. These three conditions are identical except in their description of the type of military action to be undertaken. In one condition, respondents read about a planned drone strike; in another, about an attack by piloted aircraft; and in third, about the use of American ground troops.

Survey experiments such as this, which form the core of our empirical analysis, have two key advantages for answering the research questions motivating this book.[66] First, their conditions are designed to answer specific research questions. Unlike the survey research about the Islamic State, the experiment described above includes conditions for the three types of military force that social science theory suggests should produce different patterns of responses. The foundation of this approach is the product of a great deal of research that finds that respondents' attitudes and opinions are heavily influenced by the context in which survey questions are answered.[67] In our survey experiments, this context is the information included in the conditions, such as the type of military force employed. Second, survey experiments allow us to conclude that such differences are causal rather than coincidental. Since respondents are randomly assigned to conditions, it is reasonable to conclude that any differences we find across conditions are due to their content rather than some other factor.

We recruited respondents for our survey experiments from two sources. The commercial survey firm YouGov provided respondents for two experiments. YouGov maintains a large panel of individuals willing to take surveys in exchange for points that can be redeemed for gift certificates and other items. Since individuals willing to complete surveys may not be representative of the adult population in the United States, YouGov matches samples drawn from its panel with samples drawn from the population at large, using probability sampling based on a range of individual characteristics such as age, partisan identification, and other political and demographic characteristics. The YouGov approach to panel recruitment and matching is now widely used in research on public opinion.[68]

For the remainder of our experiments, we recruited respondents from Amazon's Mechanical Turk, an online labor market where individuals complete tasks, including surveys, in exchange for small payments. Mechanical Turk has quickly become one of the most common sources of respondents for online experiments in the social sciences because of its flexibility and low cost compared to more representative samples. It differs from online respondent panels provided by firms such as YouGov in one important respect. YouGov panels are designed to maximize the similarity between samples and a larger population, such as adults in the United States. In contrast, Mechanical Turk samples differ from the adult population in systematic ways; respondents are typically younger and more politically liberal, for example.[69] This difference raises an important question: can we draw conclusions about how Americans in general think about and evaluate combat drones when using Mechanical Turk survey experiments? In other words, do the characteristics that distinguish our Mechanical Turk respondents make them likely to respond differently to experimental treatments?

This is an important concern, which we address in a number of ways. The first is to acknowledge that the primary goal of this book is to understand how a new military technology, combat drones, alters what existing theories tell us about the public's support for the use of force. In other words, the primary goal of our experiments is to test theory. Experiments are the ideal tool for achieving this goal, because they provide the researcher the ability to control conditions and treatment assignment, which permits one to conclude that any findings are most likely caused by random assignment to different conditions, rather than by other, potentially confounding factors. At the same time, we do think there is some scope to conclude that the experimental effects we report are likely to occur not just among members of our Mechanical Turk convenience sample, but also among other samples. The similarity of experimental results conducted with probability and convenience samples has been subject to a large body of research in recent years, and much of this work finds that conditions produce similar effects across these types of samples,[70] especially when relevant covariates are included in statistical models.[71] We build on this line of work as well; in chapter 3, we report the results of identical survey experiments on our probability and convenience samples conducted within a few weeks of each other and find that these two groups of respondents reveal qualitatively similar results. Although our work is not designed primarily to produce estimates of how the population at large would respond to our experimental conditions, these factors do suggest that future research could begin with the expectation that there would be a close correspondence between probability and convenience samples on this topic.

Our research approach is designed to maximize the transparency of the data and process of analysis we use in reaching our results. All of the data and analysis code is available for other researchers to inspect and to use in their own work.[72] We also preregistered the research design for the two experiments that relied on a nationally representative sample of respondents and one experiment that recruited respondents from Mechanical Turk.[73] Prior to receiving the data from these experiments, we pre-registered our plan for using this data to test hypotheses about how drone technology influences support for the use of force. Pre-registration is, among other things, a form of committing in advance to our hypotheses. By declaring in advance the data we plan to collect, how many respondents would be included, and how we would analyze data from our experiments, it reduces the incentive to conduct analyses after the fact in a manner that is consistent with our hypotheses. This should increase readers' confidence that the results reported from these experiments have not been subtly shaped by pressure to report novel or statistically significant analyses, and that they are based on the implications of drone technology that we deduce from existing theories of public opinion rather than post hoc theorizing.[74]

Outline of the Book

Chapter 2 begins our empirical investigation, using a survey experiment with a representative sample of Americans to answer two questions. First, do attacks with drones receive more support than other attack types, such as the use of ground troops or air power? Second, how large is the effect of drones compared to other influences on support for the use of force? These questions are at the core of the debate about drones, but existing research has not addressed them directly. This chapter takes on this task, using a conjoint survey experiment that permits comparison of the effects of drones to many other factors that shape attitudes toward military action. It finds that attacks from drones do garner more support than other attack types, but that the size of this effect is moderate, roughly equal to other factors identified by previous research such as gender and the likelihood of military success.

Chapter 3 addresses one of the biggest potential benefits of and potentially serious objections against using drones: the possibility that pilot invulnerability may increase the incidence of war by allowing politicians to fight without provoking backlash from casualty-averse constituents.

The reasoning here is that states armed with drones would have few inhibitions against using them in combat when they would not sustain any military casualties—a result that could lead not only to more wars but also to a decline in democratic accountability. We discuss an experiment with a respondent pool representative of adults in the United States showing that attacks using drones, compared to manned aircraft or ground forces, do garner more support for the use of force and that this relationship is influenced by the fact that drones minimize the chance of military casualties. However, the shift in support is fairly small and is not the only factor that shapes attitudes about violent conflict. This indicates that drones are unlikely to cause a radical increase in support for military operations, and that their effects on the incidence of war have to be understood alongside many other considerations.

In chapter 4 we consider how support for the use of force is influenced by the type of policy objective being pursued. We show that public opinion is not only sensitive to the anticipated numbers of military casualties, but also to the goals and objectives of military force. Regardless of the weapons being used, counterterrorism operations receive the greatest support, followed by foreign policy restraint, internal political change, and humanitarian intervention. This demonstrates that the perceived urgency of a threat or perceived legitimacy of a mission exert a powerful influence on support for war, which is comparable to the influence of using drones rather than ground forces or manned aircraft. When considered alongside the data presented in chapter 3, this result provides grounds for developing a more complex understanding of how drones affect the support for the use of force than the common prediction that drones are apt to simply lower inhibitions against fighting.

Chapter 5 evaluates the argument that, because they lower the costs of conflict, drones create incentives for political leaders to engage in conflicts that have little chance of success. We hypothesize that people might be more willing to support drone strikes even when there are viable alternatives to military action, since the costs of mission failure (for example in terms of military casualties) would seemingly be low. If that is the case, it would have profound implications for when authorities can use force with the support of the public. It could free politicians to initiate conflicts that have a lower probability of success while also limiting the backlash they may face when an operation goes badly. The results of this experiment suggest that this is not the case, and that respondents continue to hesitate to support military action even when drones are available.

Chapter 6 argues that the absence of military casualties will increase attention to civilians harmed by drones. Drawing on insights from psychology about "counterfactual thinking," we theorize that selectivity raises expectations about the results that will obtain when selective weapons such as drones are employed. Drones make it easier to avoid inflicting civilian casualties, and the public adjusts its expectations with this in mind. A first survey experiment tests these conjectures about how drone weaponry makes individuals more sensitive to civilian harm. Respondents primed to expect fewer civilian casualties expressed more regret, more sympathy with victims' families, and less satisfaction than did those primed with a higher risk of civilian deaths, even though the actual outcomes across treatments were identical. This effect weakens when precision strikes are carried out by manned platforms. We also argue that people expect their leaders to exploit the selectivity that drones create and to plan their attacks in ways that minimize the risk of killing civilians, especially when doing so does not reduce their ability to achieve military objectives. A second survey experiment assesses this expectation and finds that support for the use of force declines when attacks place civilians in harm's way unnecessarily. This effect strengthens when respondents are informed that the attack violates international law, suggesting that legal and moral norms exert an independent influence on assessments of the use of force. Importantly, the differences remain even when the group targeted with an attack is described as directly threatening the United States, indicating that individuals are not willing to "trade" the risk of civilian death for greater security.

The concluding chapter analyzes the implications of the findings reported in the book for the ethics of the use of force. By removing the possibility of American military casualties and appearing to be very selective, drone technology may tempt leaders to ignore the ethical injunctions of just war theory and instead to use drones to achieve quick and low-cost military successes. The empirical results of the book can help to inform this debate by providing evidence about the extent to which the American public values avoiding military casualties.

CHAPTER 2

Drones and Support for the Use of Force

The central questions we address in this book are: how do armed drones influence support for the use of military force among citizens? To what extent are the factors that have previously been identified as influences on public approval for war salient in an era when the costs of conflict appear to have been radically disrupted by drones? Our goal is to better comprehend how drones alter the American public's support for military operations, and to use this as the basis for understanding American foreign policy as this type of weapons platform plays an increasingly important role.

This chapter focuses on the first question, seeking to determine if and how much drones alter support for the use of force. We begin by assessing arguments about how the availability of combat drones reduces the domestic political costs of conflict. Lower costs for conflict could lead the United States to engage in wars that are counterproductive or that have a questionable moral and ethical basis. While these arguments make plausible inferences about how drones could influence support for the use of force, they have not been assessed in a rigorous way. There is a large and diverse literature on public opinion and foreign policy that includes analyses of previous American wars and experiments based on hypothetical conflicts. This work identifies a range of factors, such as the likelihood of military casualties and of success on the battlefield, that affect opinions about military operations. But because combat drones have only been used for about a decade, most of this rich body of work has not grappled with their implications.

This chapter brings together these two streams of work by comparing the effect of drones to these known influences on Americans' attitudes, allowing us to gauge the relative impact of this new military technology. To achieve this objective, we introduce our "attack type" survey experiment, variants of which are also employed in the next three chapters. In these experiments, respondents are presented with information about a planned use of military force, which is described as being carried out by a drone, a manned aircraft, or ground troops. The specific type of experiment we report in this chapter, known as a conjoint survey experiment, is ideally suited to comparing the effect of the attack type on support for the use of force with other, well-established influences on the public's support for military action.

This experiment yields a number of findings that shed light on how drones influence attitudes toward war. Our first and most important finding is that attacks with drones receive more support than attacks with ground troops. Drones do increase citizens' support for the use of force. Second, attacks from drones are only preferred to the use of ground troops. Respondents in our experiments expressed statistically indistinguishable levels of support for drone strikes and air strikes. Third, this relationship holds for both Democrats and Republicans. Previous research indicates that Republicans are generally more favorably disposed to military action. Consistent with this argument, our experiment finds that when asked to assess the desirability of any type of military action, Democrats provide lower levels of support than do Republicans. However, when asked to compare the desirability of drone strikes and the use of ground troops, both Democrats and Republicans prefer the former. This suggests that there is a cross-party consensus favoring the use of drones, unlike much of domestic and foreign policy in the contemporary United States, and that changes in the partisan makeup of Congress and the presidency are unlikely to restrain the use of this technology.

We also find that drones have a substantive effect on attitudes toward conflict that is of the same magnitude as other, well-established influences on support for the use of force. For example, attacks carried out by drones increase support over those conducted with ground troops by about as much as do attacks that are likely to succeed or that are unlikely to cause civilian casualties. Drones, then, noticeably increase support for the use of force. In this and the next chapter, we suggest that it is pilot invulnerability, rather than selectivity, that has the most substantial influence on public opinion about using drones. Respondents in our survey experiments foresee a far lower risk of military casualties when drones are employed than they do when attacks are carried out by strike aircraft or by ground troops.

The following chapters build on these findings. In the next chapter, we explore the role of invulnerability and casualty aversion in more detail. Subsequent chapters analyze how drone technology relates to other influences on public opinion. We ask, for example, do drone strikes receive greater support across the range of objectives that can be achieved with military force? Do the lower costs of drones make individuals more willing to endorse risky military missions (chapter 4)? Do they alter how people think about civilian casualties (chapter 5)? The combination of a comparative perspective in this chapter, along with analysis in later chapters of how drones may modify other influences on support for the use of force, allows a thorough assessment of the political consequences of this new military technology.

Drones and Support for Military Action

There are two reasons to think that attacks by drones could receive more support than attacks with manned aircraft or ground combat personnel. The first is the invulnerability that drones provide to their operators. One of the most widely recognized constraints limiting the United States' exercise of its military strength are political leaders' concerns about the public's aversion to casualties. By some accounts, Americans are highly sensitive to military casualties and quickly become disillusioned with a war as casualties mount.[1] Similarly, many commentators attribute American reluctance to deploy its full might in small wars to a fear that the American public is unwilling to bear the costs of these conflicts.[2] The extent of casualty aversion is debatable, with some studies finding that it has a small effect, that it influences elites more than the general public, or that it only operates in conjunction with other factors.[3] Nevertheless, even those who doubt that casualty aversion is a decisive factor typically credit it with having some role in dampening support for war. Without operators aboard, drones could decisively alter calculations about when to fight, making it easier for the United States to support its allies, deter rivals, and participate in humanitarian interventions.

Second, drones may be preferred over other types of military force if they are perceived as being more likely to achieve military objectives. Armed drones have only recently entered the American military's arsenal and have been deployed primarily to counter terrorist and insurgent groups. As we saw in chapter 1, a key problem in this type of attack is distinguishing combatants from noncombatants. The selectivity of drones should reduce the chance that military force will miss its target or harm noncombatants, both of which dampen support for engaging in military action. Drones may even make it possible to avoid harming low-ranking enemy personnel who may be ineffective or unwilling combatants once their leaders are killed.[4] The technical capabilities of drones may make them particularly effective in solving this problem of target selection, leading them to take on the image of high-tech super-weapons, much as precision-guided munitions did following the First Gulf War.[5]

Politicians, members of the military, and media commentators have encouraged this perception by making lofty promises about what drones are able to accomplish. The Obama administration regularly publicized the results of strikes that killed key terrorist leaders and credited these with crippling organizations that were determined to attack the United States.[6] If members of the public see drones as being more selective than manned aircraft and ground forces, then they may be more willing to support drone strikes. Drones may therefore offer the attractive promise of war that satisfies the *jus in bello* principles of discrimination and proportionality more effectively than ever before. This is particularly true when drones are compared to manned aircraft, which were used in some of the clearest violations of those norms throughout twentieth-century wars.[7]

By reducing public concern about, or attention to, the costs of conflict, drone warfare could have dangerous consequences. Protecting American soldiers from enemy fire by physically removing them from the battlefield and replacing them with drones could make it easier for the United States to use military force when it is justified. But it could also facilitate aggressive or unjustified wars. Lowering this cost of war could remove an invaluable political constraint on reckless and unnecessary conflicts, making it possible for leaders to initiate wars more easily and to avoid being removed from office if the wars go badly. Some opponents of unmanned weapons platforms even go so far as to argue that the ability to fight without suffering casualties will dramatically lower the threshold for initiating wars and promote public disengagement from use of force decisions.[8]

Even if drones are more effective than other weapons systems for certain missions, confidence in their selectivity may lead to a dangerous over-reliance on these weapon systems. Of particular concern is the possibility that they could create a "moral hazard" by lowering the costs of fighting to such an extent that the American public may be enthusiastic about using force even when victory is unlikely.[9] The ultimate risk here is that drones could cause a surge in the incidence of unnecessary wars of choice that would kill people and destroy infrastructure while delivering few strategic benefits to the United States.

Critics contend that improvements in selectivity highlighted by drone proponents are largely illusory. Selectivity is gauged within a margin of error that is always fallible, especially given the chances of drones' malfunctioning, launching attacks against misidentified targets, or striking innocent bystanders in the process of killing terrorist leaders.[10] Worse still, portraying drones as being more selective and precise may paradoxically pave the way for increased violence against civilians by making it easier to build support for attacks that are carried out in populated areas. Whereas the American public may believe that massive aerial bombardments against cities would have unconscionable effects on civilians, people could arguably be lulled into a false sense of security that a series of precise attacks directed at a city would have little or no adverse effects. Added to this, there is uncertainty about the extent to which Americans care about foreign civilian casualties at all, especially if the absence of American losses make the costs of war virtually imperceptible.[11]

In much of the literature on drones, both the appealing characteristics and the potential dangers are presented with compelling theoretical explanations. But drones' influence on support for war in the United States has not been systematically tested in most of this work. With sweeping predictions of rising American military aggression, an escalation of violence around the world, and increasing civilian casualties on the one hand, or promises of more restrained, ethically sensitive, and precise attacks on the other, it is vital to have a clearer sense of whether these expectations stand up to scrutiny, a task we tackle in this chapter.

What Shapes Opinions Regarding Military Action?

One goal of this chapter is to determine whether drones influence support for the use of force. A second goal is to estimate the size of this influence.

A difficulty here is that we do not have obvious benchmarks for what constitutes a large or small effect of drone technology. Theories suggesting, for example, that drones will lead to more support for military action do not make precise predictions of the size of this effect among members of the public. Those making these claims likewise tend to treat the public as a homogeneous group that will respond to drones in fairly uniform ways, without considering the extent to which demographic characteristics and party affiliations may shape preferences. Our solution to these problems is to compare the effects of drones to other, well-established influences on public opinion.

Previous research finds that perceptions of military success, the political goals of military action, the likelihood of civilian casualties, and the opinions of allies, international organizations, and nongovernmental organizations all exercise consistent influences on attitudes toward conflict. Kreps' study of public support for drones, which is one of the few studies that analyzes these weapons platforms in particular, shows that variations in how survey questions characterize drones' effectiveness and their status under international law influence support for strikes.[12] Although she only focuses on the effect of these two considerations, the framing effects she finds demonstrate the value of developing experiments that incorporate more variables that could potentially shape public opinion. The experiment we report in this chapter is designed to compare the substantive effect of drones to these known influences. This not only builds a more comprehensive picture of the factors shaping public opinion about drones, but also helps to link our findings to public opinion research on foreign policy more broadly. Here we briefly summarize each of these. Subsequent chapters analyze how the availability of drones interacts with these factors to determine if drones alter the manner in which these factors influence public opinion. In this chapter, though, our aim is to simply compare the size of the effects for these four factors to that of drones.

Consider first how the likelihood that a military mission will achieve its goals influences attitudes regarding war. From this perspective, individuals are more apt to support the use of force when they believe it will achieve important political or military objectives. The reasoning here is straightforward: when the United States is likely to prevail in conflicts, the public concludes that the net costs of using military force—military casualties, expenditures, the risk of a long war, and so on—are modest compared to the benefits of success on the battlefield. Early work in this vein identifies the likelihood of success, along with the clarity of the mission's objectives and its stakes for the United States, as important influences on how the public evaluates war.[13] Subsequent work has extended this line of inquiry, using public opinion surveys as well as experiments to conclude that conflicts with a high chance of success are more popular and that setbacks or advances during the course of conflict exercise a negative or positive influence on public attitudes toward war.[14] Debate continues about if and how large an effect perceptions of success have on attitudes. Some suggest that such perceptions are, in large measure, either driven by respondents' political predispositions, such as their partisan identification, or are simply a substitute for the expectation that military casualties and other costs of the conflict will be low.[15] Nonetheless, there is considerable evidence that perceptions of success continue to matter even when such factors are taken into account.[16]

Less work has looked at how civilian deaths caused by American military action alter support for the use of force. Some analysis of polling data and other sources from the Second World War onwards suggest that noncombatant deaths have at most a modest influence on American citizens' calculations.[17] More recent research has found that civilian casualties influence public attitudes, particularly since the 1970s.[18] One study, for example, finds that survey questions that mention the possibility of civilian casualties reduce support for military action, and that the size of this effect was of the same magnitude as for questions that mention American military casualties.[19] A survey experiment that compares the effects of military and civilian casualties in a hypothetical war reached similar conclusions, although most respondents, when asked, responded that avoiding military casualties was more important than avoiding casualties among noncombatants.[20] There is reason to believe, then, that the risk of civilian harm can lead to meaningful reductions in Americans' willingness to endorse military action.

Another factor influencing support for the use of force are the goals, or "principal policy objectives" (PPO) that military force is intended to achieve. Work in this vein has distinguished between four distinct goals: foreign policy restraint, which aims to deter or restrain another state from taking threatening actions; internal political change, in which force is used to assist an ally facing rebellion or other threats to its hold on power; humanitarian intervention to prevent or stop conflicts inflicting large-scale harm on noncombatants; and counterterrorism, such as using force to attack terrorist bases overseas.[21] Initial work found that foreign policy restraint receives much higher levels of support than does internal political change. Subsequent studies found that support for humanitarian interventions tends to fall somewhere between foreign policy restraint and internal political change. Humanitarian interventions have lower levels of support than the former because they are not seen as being as necessary for the protection of national security, but they are more popular than internal political change missions that lack the moral status of humanitarian operations.[22]

After the terrorist attacks on the United States on September 11, 2001, and the United States' involvement in wars in Afghanistan and Iraq, other researchers gauged support for counterterrorism missions, which received the broadest level of support.[23] As with other findings on public opinion and foreign policy, these have not gone unchallenged. Some research suggests that how conflicts are framed as PPOs is flexible and, to some degree, under the control of political leaders. Individuals who lack information about foreign policy, in particular, may be susceptible to such framing and follow the judgments of leaders with interests and ideologies similar to their own.[24] Nevertheless, there is considerable evidence that PPOs are important to the public, even if some individuals are influenced by how the conflicts are cast by political leaders rather than by the reality on the ground.

Finally, we compare the effect of drones on public attitudes to the effect of opinions of international actors including allies of the United States, international organizations, and nongovernment organizations like transnational human rights groups. Recent research suggests that these actors can influence Americans' attitudes under certain conditions. The starting point for this approach is that many individuals have general preferences or pre-dispositions for the goals and means of American foreign policy, but lack the detailed knowledge of the international environment that would allow them to translate these into support for or opposition to specific policy measures, including drone strikes.[25] These individuals can use the opinions of international actors with expertise in such policies as a signal for how to translate their general preferences into positions on particular policies. In doing so, individuals compare their general predispositions to those of the international actors. If these are in broad agreement, the individual can adopt the specific policy preference of the better-informed international actor. Disagreement between the two is also informative; in such cases, the individual can oppose policies that are supported by the international actor.[26]

Comparing Influences on Support for the Use of Force

In this chapter we report the results from an experiment designed to provide a comparative perspective on attitudes about the use of force, before turning to the more specific issues later in the book. We developed a fully randomized choice-based conjoint experiment,[27] designed specifically to assess the influence of many factors simultaneously and to compare their relative impact on attitudes about military action.[28] Respondents were a representative sample of 1,000 adults in the United States recruited by the survey research firm YouGov.[29] Each respondent first read the following introduction to the experiment, placing them in the role of evaluating two attacks by American forces:

> For the next few minutes, imagine you are the Secretary of Defense of the United States. Military officials have provided you with intelligence assessments about two attacks that the United States might carry out in the near future. For each pair of proposed attacks, indicate which you personally would prefer to authorize. In making your choices, remember that military resources are limited, and the United States can only carry out some attacks that are proposed by military planners. Even if you are not entirely sure, please indicate which of the two you prefer.

Respondents were then presented with six attributes of each attack, each of which can be described in multiple ways. Table 2.1 lists the levels of the attributes included in the experiment; those in bold serve as the baseline treatments in the statistical analysis reported below. The levels of each attribute were randomly selected for each of the two attacks that respondents were asked to compare. This allows us to estimate the effect of each attribute value on support for the attack. In the analysis below, each value is measured on the same scale, allowing direct comparison of their causal effect on the dependent variable. Each respondent makes five comparisons of two attack plans, each on a separate screen, resulting in 10 responses for each variable and a total of 10,000 observations. Table 2.2 depicts one such comparison between two attacks.

As can be seen in table 2.1, this conjoint experiment allows for the simultaneous inclusion of many attributes. We did not include military casualties as an attribute in this experiment. One reason for this is that drone strikes would produce no military casualties. To account for this, the experiment would need to constrain the value of the military casualties attribute to equal zero in all cases where the attack type was described as a drone strike. This means that the assignment of attribute values would not be fully random, potentially weakening the experimental design. While there are statistical techniques that account for such deviations from a truly random design,[30] our decision to forgo including this attribute was motivated by a more fundamental consideration. Had we not done so, every attack described as a drone strike would be accompanied by information that this would produce zero military casualties, while the other attack types—air strikes and ground troops—would always describe the expected number of military casualties as zero or greater. We worried that always comparing drones with no casualties against air strikes or attacks by ground troops with the possibility of casualties would lead some respondents to draw close attention to this pattern, and base a disproportionate share of their choices on the presence or absence of military casualties rather than on other factors.

Nonetheless, we are interested in how perceptions of military casualties shape attitudes. To address this, we followed up the conjoint experiment with questions that asked respondents to estimate the likelihood of military casualties resulting from a drone strike, a strike with manned aircraft, and the use of ground troops. Based on the literature on military casualty aversion, we expect that the likelihood of military casualties foreseen by our respondents would be lower when the attack is described as coming from a drone than from aircraft or ground troops.

One difference from the experiments discussed in subsequent chapters is that the descriptions of the attributes are shorter in this conjoint experiment. Short attribute levels should reduce the chance that respondents might become confused or fatigued and stop the survey before completing it. They also allow us to assess the degree to which small changes in how an attack is described influence respondents' attitudes. In the next chapter, experimental conditions for attacks carried out by drones and by aircraft included information that these attacks would have a low risk of military casualties. This information about military casualties is not included in the conjoint experiment, allowing us to assess whether this leads to substantive changes in how attack type influences support for the use of force.

To analyze the data, we first generated indicator variables for each level of each attribute. We then set one level as the baseline for each attribute. The effect of each attribute is measured as the change in the probability of preferring an attack compared to this baseline. For example, the baseline for the attack type attribute is an attack described as a drone strike. The effect of attacks described as air strikes and ground troops is the difference in the probability of preferring these attacks from those described as drone strikes. These estimates are calculated as average marginal component effects (AMCEs), which identifies the average differences of being preferred when comparing two attribute values over all of the possible combinations of other attribute values, with standard errors clustered on respondents.[31]

Figure 2.1 summarizes the results of this exercise. We have omitted information about international opinion to make the figure easier to interpret; full statistical results are presented in the chapter appendix. The levels in bold with dots at zero on the vertical line are the baseline levels for each attribute. The horizontal axis indicates how treatment by each

TABLE 2.1. Attributes and Values for Conjoint Experiment

Attributes	Values
Type of attack	**Missiles fired from unmanned drone aircraft** Missiles fired from military aircraft with a crew of two officers Ground troops in armored vehicles
Target of attack	**Group of low-level enemy forces** Commander of enemy forces
Likelihood that attack will achieve its objective	**Low** Moderate High
Likelihood of civilian casualties	**Low** Moderate High
Objective of attack	**Prevent a foreign country from providing safe haven to al Qaeda terrorist bases** Prevent a foreign country from engaging in genocide and ethnic cleansing Prevent a foreign country from disrupting oil shipments from the Persian Gulf
International opinion	**Supported by NATO allies in Europe** Opposed by NATO allies in Europe Supported by the United Nations Security Council Opposed by the United Nations Security Council Supported by international human rights groups Opposed by international human rights groups

Note: Baseline attribute values are indicated in bold.

attribute level reduces or increases support for an attack compared to the baseline level. Dots indicate point estimates, and horizontal lines identify the 95 percent confidence intervals for the ACME of each attribute level. For example, attacks with air strikes receive only slightly less support than drone strikes, and the difference between these two estimates is not statistically significant.[32] But when the attack is carried out by ground troops, the probability of preferring the attack declines compared to drone strike. Here it seems that the greater risk of losing American soldiers in ground combat operations encourages respondents to become more cautious about endorsing military action, a point we return to below and in chapter 3. The most likely explanation is that respondents see aircraft as having a decisive effect on protecting American military personnel, such that the presence or absence of an onboard crew matters little when it comes to trigger-

TABLE 2.2. Comparison of Two Attacks

Please carefully review the information about two attacks detailed below, then answer the questions.

	Attack 1	Attack 2
Likelihood that attack will achieve its objective	Low	Low
International opinion	Supported by the United Nations Security Council	Opposed by NATO allies in Europe
Objective of attack	Prevent a foreign country from engaging in genocide and ethnic cleansing	Prevent a foreign country from providing safe haven to al Qaeda terrorist bases
Target of attack	Commander of enemy forces	Group of low-level enemy forces
Type of attack	Ground troops in armored vehicles	Missiles fired from military aircraft with a crew of two officers
Likelihood of civilian casualties	High	Moderate

Question: Which attack do you prefer?

☐ Attack 1
☐ Attack 2

Question: On a scale from 1 to 7, where 1 indicates that the United States should definitely not carry out the attack and 7 indicates that the United States definitely should carry out the attack, how would you rate attack 1?

Question: On a scale from 1 to 7, where 1 indicates that the United States should definitely not carry out the attack and 7 indicates that the United States definitely should carry out the attack, how would you rate attack 2?

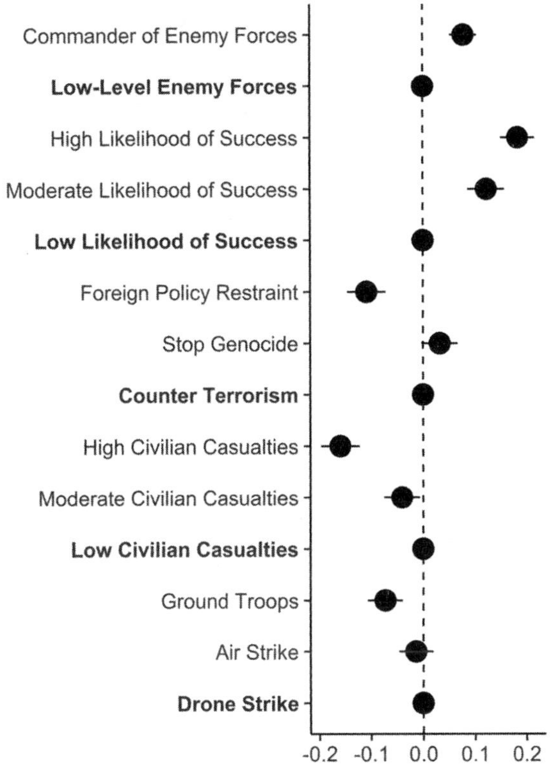

Fig. 2.1. Effects of attributes on probability of preferring attack

ing risk aversion. This attitude reflects the country's ongoing participation in asymmetric wars against enemies that have little anti-aircraft capacity. One potential concern is that this perception of pilot invulnerability may persist into conventional wars to create false expectations of low casualties regardless of the type of aircraft involved and the enemy's capacities.

Considerable evidence suggests that Republicans are more likely to endorse the use of military force than are Democrats.[33] This is the case in our conjoint experiment. Recall that in addition to asking respondents to indicate which of two attacks they prefer, we also asked them to rate their support for each attack. This seven-point rating ranges from 1 to 7, with higher values indicating the respondent expressed a greater willingness to authorize the attack. Figure 2.2 summarizes the distribution of these ratings for all the attack plans presented in the experiment for Democratic and Republican respondents. Compared to the ratings of Democratic re-

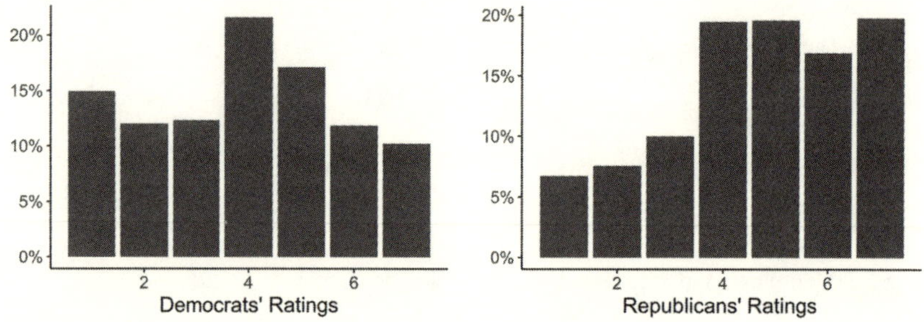

Fig. 2.2. Partisanship and ratings of attacks

spondents, those of Republicans are skewed to the right, with over half of the ratings in the three highest values for this variable. While the average rating by Democratic respondents was 3.9 on the seven-point scale, the average for Republicans was 4.7, and this difference between partisans is statistically significant ($p < .01$). Our Democratic respondents, then, were less willing to authorize attacks.[34]

However, these differences in willingness to endorse the use of force in general disappear when we consider specific attack types. We twice repeated the conjoint analysis used to create figure 2.2 but limited the sample to Democrats and Republicans. The results of this exercise are depicted in figure 2.3, which shows the effect of each attack type for supporters of each party on their willingness to prefer one attack over the other. (To simplify the display of information, this figure omits the other attributes; details can be found in models 3 and 4 of table A2.1 in the chapter appendix.) Democrats and Republicans evaluate attacks in similar ways. Both prefer drone strikes to the use of ground troops. Republicans also prefer drone strikes to air strikes, although this difference does not meet the conventional threshold for statistical significance.

The absence of partisan differences in support for drone strikes over the use of ground troops explains, in part, the continuity in the use of drones. Drone strikes against militants outside of traditional combat zones began under Republican president George W. Bush. President Bush accelerated the pace of drone strikes significantly in early 2008 and loosened the conditions under which they could be used to target militants. Early drone strikes were authorized when a target was positively identified as a militant. Beginning in 2008, drone strikes also targeted individuals or groups that

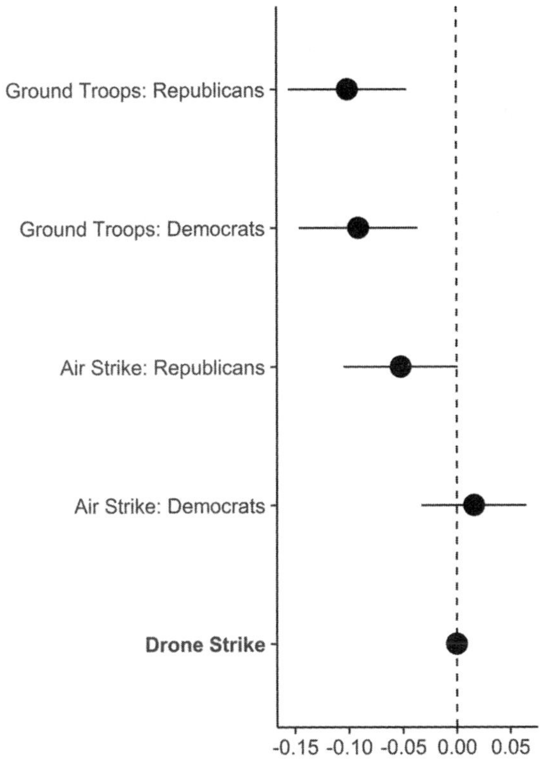

Fig. 2.3. Effects of attributes on probability of preferring attack by party

exhibited the "signature" of militants, such as a group of armed men traveling in a convoy or residing in a building that had been used by militants in the past. Barack Obama was elected on a promise to extract American troops from wars in Afghanistan and Iraq. But this aversion to large-scale combat did not lead his administration to curtail drone strikes; instead, the number of strikes expanded dramatically during his first term in office. Furthermore, the use of drones by both Presidents Bush and Obama evinced little criticism or even public oversight by members of Congress, even when the opposing party controlled the legislature. At both the mass and elite level, then, there is a consensus across party lines favoring drone strikes.[35]

Attacks in which the chance of civilian casualties were moderate or high receive less support than those with a low likelihood of civilian harm. Concern about inflicting civilian casualties does limit support for the use of force. The same pattern exists for mission success; attacks described as having a moderate or high chance of achieving their objectives are more likely to be preferred. This coincides with Kreps' study of the influence of question framing on responses to opinion polls about drones, as her findings show a drop in support when questions indicated a high likelihood of harming civilians or inflicting disproportionate damage.[36]

At first glance, this may appear to be an obvious result. After all, it is clearly more advantageous to launch attacks that are apt to succeed than those that are less likely to succeed, and previous research has found that anticipations of success play a central role in determining overall levels of support for war.[37] However, this result takes on special significance when drones are involved because of the debate over how the costs of fighting may be displaced onto civilians. Critics of drones have worried that the low costs of launching strikes could induce the United States to wage wars even when the chances of winning are low. This might result in greater willingness to escalate a conflict, less patience with diplomatic alternatives to fighting, and a proliferation of wars that are unlikely to yield any redeeming benefits—all while placing civilians at greater risk than if there were stronger inhibitions against fighting. Kreps says that her findings support these fears, as she concludes that most polls downplay the potential consequences of drone warfare to create a misleadingly optimistic perspective on the strikes. This is an issue we explore in greater depth in chapter 5, with additional testing and a discussion of the various causal mechanisms that might arguably create such a moral hazard.

Compared to attacks with the objective of countering terrorism, those with the goal of restraining a foreign state's aggression receive less support, while those with the objective of stopping genocide receive more support (although this difference from counterterrorism missions is not statistically significant). We will revisit this ranking of preferences in chapter 4, where we find that counterterrorism operations receive more support than alternative mission types. We also explore some of the possible reasons for this variance in mission preferences across attack types. For now, it is important to bear in mind that preferences here are not as consistent as some of the other patterns we identify, which is likely due to changes in what threats seem most urgent when the experiments are conducted and how each type of mission is framed when it is described in a hypothetical attack scenario. The high support for counterterrorism compared to restraining foreign aggression probably has much to do with how threats have been framed since 9/11. Political and media elites have consistently emphasized the threat that terrorism poses and used this as a rationale for launching drone strikes. This means that it is a pattern that will change along with shifts in the security landscape and elites' framing of threats.

Although the relative importance of counterterrorism is likely temporary, it does raise some concerns for the country's ability to respond to security threats. By many accounts, politicians and members of the media dramatically overstate the threat of terrorism.[38] The comparatively high support for drones may be a consequence of this threat inflation, which calls the value of these preferences into question. Preoccupation with counterterrorism could present difficulties for leaders seeking to reorient drone operations to contend with conventional military threats, though as we pointed out in the previous chapter, drones may be less effective in conventional operations.[39] Similarly, it is possible that support for drones in any context will decline if terrorism is framed as a less urgent threat. Experimental designs cannot account for these long-term trends in attitudes toward drone use, so it will be important to revisit this problem if perceived national security priorities change.

Attacks that target commanders of enemy forces receive more support than those against low-level fighters. This is consistent with what we should expect from how drone strikes and other forms of military action have been framed throughout the War on Terror. Attacks are typically described as being made against high-ranking terrorist leaders who are uncompromising enemies determined to kill innocent people.

The most puzzling pattern concerns the influence of preferences of the international community. Predictably, attacks opposed by the United States' NATO allies are less likely to be preferred than those in which NATO allies support the attack, the baseline category for this attribute. But support <u>or</u> opposition to the attack by either the United Nations Security Council or by human rights nongovernmental organizations are less likely to receive support. This is surprising, as existing research would suggest that these factors should have opposite effects. Kreps finds that references to international humanitarian law (IHL) can influence support for drone strikes in public opinion polls, with support declining as questions are framed to suggest that attacks may be legally problematic. Although she focuses on IHL, rather than on international organizations, her results suggest that the pressure of norms coming from the international community should have some constraining influence on public opinion.

Separate analyses of Democrats and Republicans shed some light on this paradoxical pattern of support for attacks. Among Democrats, the opinions of international actors has a consistent effect on support for the use of force. Opposition to an attack by NATO allies, human rights groups, and the United Nations Security Council all lead to less support for an attack among Democratic respondents in our experiment. Republicans were less likely to support an attack if it was opposed by NATO allies or endorsed or opposed by the Security Council or by human rights nongovernmental organizations. This result indicates that Republicans in particular are wary of international organizations, especially those that are not dominated by the United States as thoroughly as NATO or that have previously acted as checks on US military operations (as the Security Council has). This aligns with previous findings that have shown that Republicans tend to dislike multilateralism except when it is clearly in American interests and that Democrats are more inclined to see an inherent value in fostering international cooperation.[40] It also coincides with the parties' traditional strategies, as leading Republicans are more inclined to denigrate multilateralism and assert American independence in foreign policy matters.[41]

Figure 2.1 also allows comparison of the relative size of the influence of drone strikes to the other factors that influence support for the use of force.[42] The difference between the effect of an attack carried out by ground troops compared to a drone strike is of roughly the same magnitude as the difference between attacks with the objectives of counterterrorism and foreign policy restraint, and it is smaller than attacks with a high probability of leading to civilian casualties and those with a moderate and high chance of success. This is borne out by our findings in chapters 3 and 4, where we also conclude that while the availability of combat drones does increase the public's willingness to condone military action, the substantive effect is relatively modest in size. Other factors, especially the political objective and chance of success and civilian harm, continue to exert sizable effects on respondents' calculations even when accounting for drone technology. This comparative perspective on the magnitude of influence is one reason why it is helpful to start with this experiment before delving into the more focused analysis of specific variables.

Another way of thinking about this is in terms of whether using drones rather than ground forces would realistically cause a decisive shift in opinion toward war. When judged by this standard, the effect of using drones is not large. Policymakers may be able to increase support for military ventures if they use drones or piloted aircraft rather than ground forces, but the difference between the aerial and ground-based approaches shows that this change in the type of military force involved will only tip the balance for or against wars in instances if opinion is closely divided. When a clear majority favors or opposes fighting, the choice of means may not matter a great deal. Despite being extremely controversial, the American invasion of Iraq in 2003 was consistently favored by more than 70 percent of the population.[43] With such enthusiasm, the Bush administration faced no need to make more extensive use of drones to lower the costs of fighting. The war was much less popular in the United Kingdom, but even there polls found support exceeding 50 percent.[44] Here again it was not necessary to increase reliance on drones even when there was a much clearer division between the pro- and antiwar camps.

Recall that this experiment did not include military casualties as an attribute so as not to suggest too strongly to respondents that concern about such casualties should drive their evaluations of the attack. As we will discuss in more detail in chapter 3, there is good reason to believe that avoiding casualties is an important reason for preferring drone strikes to other attack types. To determine if respondents in this experiment viewed drones as reducing the risk of military casualties, we asked them to estimate the likelihood of harm to military personnel from attacks carried out by drone strikes, air strikes, and ground troops. Responses were on a seven-point scale, ranging from estimates that casualties would be extremely unlikely to extremely likely, which we collapse into three categories: likely, neither likely nor unlikely, and unlikely. This question was posed to respondents after they had completed rating the ten attack plans in the conjoint experiment. Results are summarized in figure 2.4. Air strikes and ground troops were viewed as placing military personnel markedly more at risk of harm than were drone strikes. This is consistent with the idea that an important advantage of drones is that they minimize the risk to American soldiers.[45]

It is worth noting that although respondents rated the chance of American military casualties following from drone strikes much less likely than sustaining casualties in ground operations or strikes from manned aircraft, some indicated casualties were possible. If one of the foremost advantages and potential dangers of employing drones is that they can circumvent sensitivity to military casualties, then these effects will only be realized if the general public anticipates drones will lead to casualty-free wars. However, the findings presented in figure 2.4 indicate that respondents are not

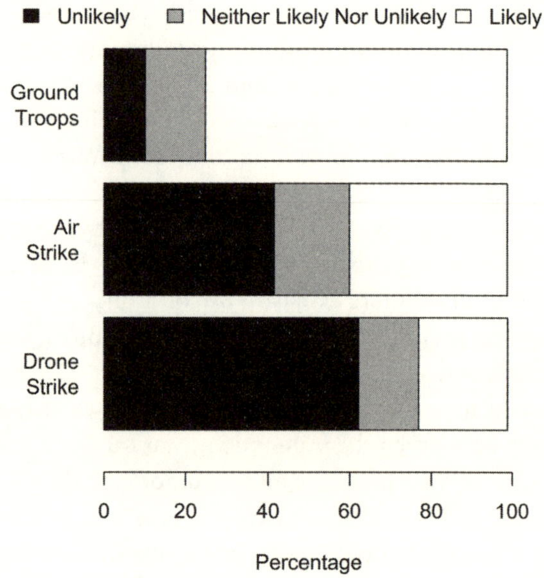

Fig. 2.4. Estimates of military casualties by attack type

willing to entirely discount the chance of military casualties, even when they read vignettes that describe an attack being conducted exclusively with drones and are not given any indication that enemies have a chance to retaliate.

Conclusion

Many of the findings we discussed in this chapter are probed further with the experiments we present later in the book. The level of agreement between multiple experiments using different designs, involving different respondent pools, and with representative and nonrepresentative samples helps to increase confidence that any particular finding is not an aberration and shows the experiments can be taken as revealing the underlying forces driving attitudes toward drone strikes.

Aside from those we have already mentioned, one of the most important points of agreement across these experiments is that air strikes receive nearly the same levels of support as drones. The narrow gap between these two options for using military force indicates that the presence or absence of a pilot onboard an aircraft makes relatively little difference in determining whether the American public will support military operations. The fact that the difference between manned aircraft and drones was not statistically significant in the conjoint experiment is especially important because this is a more representative sample of the population. In later chapters we will employ respondents recruited from Mechanical Turk. In some of these experiments we find a larger gap between support for manned aircraft and drones, though it still remains small when judged in terms of whether this difference could influence policy choices. But a consistent finding across all our experiments is that attacks with ground troops receive less support than those carried out by drones.

These results help us to better understand how drone technology could alter the domestic politics of conflict. They are consistent with the idea that policymakers will be able to maintain higher levels of support for conflicts when they avoid committing ground forces. The choice between drones and manned aircraft is more ambiguous. Leaders could be able to achieve slightly higher levels of support for missions by employing drones, especially because this could eliminate the chance of American soldiers being injured or killed. But using drones does come at a potential cost of fueling the ongoing debate about what many see as an excessive reliance on these weapons platforms. The results also show that the mission type is an important determinant of an attack's popularity. Counterterrorism operations were more popular than humanitarian interventions or attacks against foreign states—an understandable response given that terrorism is often presented as the most serious threat to American national security and that the other missions could result in much longer deployments of American forces overseas.

Nevertheless, the levels of approval for various mission types needs more attention for several reasons. First, in this chapter we have only tested three different principal policy objectives. The available literature has identified as many as four and has framed these in varying ways. In chapter 4 we consider some of the alternative formulations of the PPOs and include internal political change to those we investigate. Second, and even more importantly, PPOs only refer to the general type of operation and may take countless different forms depending on what is at stake under particular circumstances. Even though counterterrorism operations receive support in the scenario we described in this experiment, it is possible that counterterrorism may be less appealing when the attack conditions are changed. The same is true for each of the other PPOs. By testing our results with a different experiment in chapter 4 we can develop a clearer picture of the ranking of PPOs and confirm the perceived importance of counterterrorism operations by showing that these are consistently more popular than other mission types.

Our conjoint experiment informs respondents of the likelihood of success, characterizing it as high, moderate, or low. In some instances, this reflects the way information is presented prior to an attack. Political leaders, military elites, and media commentators frequently make promises about the likelihood of a good outcome when resorting to military force. However, there are times when these people either refrain from making any clear promises or issue contradictory claims. Disagreements between political parties are especially likely to result in divergent assessments of whether victory is possible.[46] Audiences may also be skeptical about these claims because promises of quick and easy wars so often turn out to be wrong. The information landscape in the lead-up to war is therefore usually more complex than the tripartite success prompt we include in the conjoint experiment. Even more importantly, the attack type could alter the anticipated outcome. Respondents may expect that drone strikes will be better at accomplishing the missions they read about if they see these machines as being more precise and efficient than other attack types. Alternatively, they could fear that drones are potentially counterproductive because of the controversy surrounding these weapons platforms and the risk of provoking harsh reactions from foreign populations. With these challenges in mind, we introduce another experiment in chapter 5 that asks respondents to form their own opinions about the likelihood of success. Here respondents draw inferences about the possible outcome from the details of the operation and the means of attack selected.

Another important point to consider is how preferences for attack may change depending on the presence or absence of peaceful alternatives. Like most survey experiments about use of force decisions, our conjoint experiment does not give respondents clues about whether the attack described is truly necessary. It may therefore predispose respondents to see violence as being justified or accepting the available response. A charge that is often directed against drones is that they lower the threshold for using military force, and it is impossible to determine whether this is the case if experiments push respondents to authorize some type of attack.[47] We therefore add in chapter 5 further nuance to our account of how drones influence valuations of success by considering how responses change when the experiment gives respondents details about whether nonviolent options are available. If drones really do make it easier to wage wars, then we should expect to see a preference for drones over other types of force when there are promising nonviolent alternatives.

The conjoint experiment showed that support for an attack declines as the likelihood of civilian casualties increases, but the considerations influencing civilian casualty sensitivity are potentially more complex than this direct relationship. Concern for civilians may not be a fixed characteristic. It is possible that using drones to reduce the harm to civilians could inspire heightened sensitivity to civilian casualties, thereby making the public less likely to support subsequent operations.[48] If this is the case, then reliance on drones may be cyclical, with each attack that avoids harming civilians increasing the demand for other attacks to do likewise. Alternatively, some commentators have argued that drones might cause a decline in sensitivity to civilian casualties over time. They raise the possibility that drones may cause public disengagement, encouraging audiences at home to pay less attention to the costs of war, including the suffering of foreign civilians.[49] We explore these issues in more detail in chapter 6.

CHAPTER 3

Drones, Casualties, and Attitudes

One long-standing debate in research on public opinion and foreign policy is the degree to which support for the use of force declines when military casualties are incurred. Influential works in this area conclude that in the public's mind, the risk of losing soldiers is among the most prominent and important costs of military engagement.[1] Others, though, contend that casualties are less decisive, as citizens balance the costs of such casualties against the political goals of military action and the likelihood of battlefield success.[2]

By rendering their operators invulnerable to enemy fire, drones upset the casualty calculations that have influenced decisions relating to the use of military force and have structured the academic debate over how public opinion responds to the possibility of armed conflict. Pilot invulnerability raises two important problems. First, research on casualty aversion focuses on operations involving human combatants, leaving considerable uncertainty about how military casualties influence public opinion when it is possible for belligerents to fight without exposing soldiers to any danger. Testing the effects of casualty aversion is particularly important now that the availability of drones could potentially make risks to soldiers superfluous. It is possible that drones could not only be preferred over other attack types, but that they may also alter existing attitudes toward other attack types because of their comparative advantage in minimizing risk.

Second, studies of the influence of casualty aversion tend to focus more heavily on how losses during a war compel an end to fighting. Less attention is devoted to what effect anticipated casualties could have on the public's desire to *initiate* conflicts or to expand ongoing conflicts into new areas.[3] This question of preventing new engagements is especially salient during the War on Terror, which has been characterized by the spread of military operations to new conflicts and countries, each time with questions about the permissibility of the new area of operations. As we showed in chapter 1, combat drones have been central to the rapid expansion of operations geographically, yet it remains unclear whether or not this expansion has been made politically easier by drones' pilot invulnerability.

Beyond these general questions relating to casualty aversion, there is also an urgent puzzle that gets to the heart of the issue of whether drones are justifiable weapons platforms. Drones allow the use of force while reducing the risks that military personnel face. Pilot invulnerability makes it possible to effectively wage wars at a distance, arguably without triggering any feelings of casualty aversion that the public may have. Peter Singer, whose work has done much to highlight the political implications of new military technologies, writes that:

> The strongest appeal of unmanned systems is that we don't have to send someone's son or daughter into harm's way. But when politicians can avoid the political consequences of the condolence letter—and the impact that military casualties have on voters and on the news media—they no longer treat the previously weighty matters of war and peace the same way.[4]

The absence of military casualties may lead, as Singer suggests, to more public support for the use of force when drones are employed compared to attacks that create higher risks of military casualties. If casualty aversion[5] reduces support for the use of force, then drones create a politically easier way to strike overseas. This concern has been taken up by many other critics of drone warfare and has become one of the chief arguments against using drones. From this perspective, civilians' sensitivity to casualties is a beneficial constraint on use of force decisions that helps to prevent wars, especially wars that are not essential to national security.

In this chapter we consider existing research on how military casualties influence support for wars, focusing on whether the availability of drones influences the public's willingness to initiate armed conflict. Our expectations are summarized in table 3.1. Given the evidence that the public is responsive to casualties and the concern that drones may circumvent this constraint on fighting, our first and most central conjecture is that the public will express greater support for military missions in which there

TABLE 3.1. Casualty Aversion Expectations

Expectation 3.1: The public will express greater support for prospective military missions in which there is no risk of military casualties because drones are employed compared to missions where such casualties may occur.

Expectation 3.2: Individuals expect fewer military casualties when attacks are carried out by drones then when they employ air power or ground troops.

is no risk of military casualties because drones are employed compared to missions where such casualties may occur. The results of the conjoint survey experiment in chapter 2 are consistent with this conjecture. In this chapter, we begin to try to understand *why* drones get more support than do other attack types.

Our second expectation is that casualty aversion drives this relationship. We assess these expectations with two survey experiments that systematically alter the type of attack—by drone, by combat aircraft with onboard crews, or by ground troops—being considered by American leaders, and then ask respondents how strongly they support or oppose this use of force. We find that respondents to whom the attack was described as being carried out with a drone expected far fewer military casualties, but the same likelihood of success and of civilian casualties, than do those for whom the attack was described as an air strike or as being carried out by ground forces. This suggests that casualty aversion, but not perceptions of victory or concern over the possibility of harming noncombatants, is what leads to more support for drone strikes.

Consistent with the chapter's first expectation, as well as the findings reported in chapter 2, respondents who read about a planned drone strike exhibited greater support for the attack than those reading about an attack with ground troops. We subject the relationship between attack type and support for military action to a number of additional tests, including duplicating it on different samples of respondents and controlling for a host of other factors that influence support for the use of force, to increase confidence that the finding is not due to chance but reflects concern about minimizing the risk American military personnel face during war.

Casualty Aversion and Support for War

John Mueller's study of public opinion during the wars in Korea and Vietnam did much to call attention to casualty sensitivity and illustrate its dynamics.[6] He found a similar pattern at work in both conflicts, with support for the wars declining as the number of casualties increased. The decline did not follow a consistent progression. Rather, there was a sharp drop in support following the initial losses, then comparatively smaller declines later. Mueller concludes that it takes a larger number of casualties to produce the same level of decline in support as a war progresses. This unequal perception of casualties may be the result of those whose support for fighting is weak becoming easily disillusioned and those with stronger commitments to the cause having greater tolerance for loss. The shift in the relative weight of casualties could also be linked to changes in a willingness to bear risks as a conflict develops. Whatever the exact mechanisms behind this change, the implication is that the effects of casualty aversion are apt to be especially strong early in a conflict but that wars may persist past the initial drop and slowly lose popularity as casualties mount, eventually forcing policymakers to withdraw military forces or negotiate a peace agreement when support falls too low.

Other studies of the Vietnam War echoed Mueller's findings and were supported by a more widespread sense that the war was hindered and ultimately lost because of the limited support it received.[7] Martel argues that "In Vietnam, mounting U.S. casualties contributed to the collapse of public support for the war because it fostered the perception that victory was not a realistic outcome and that intervention was not worth the cost in lives and resources."[8] This narrative of casualty aversion fueled subsequent fears of repeating the mistakes of Vietnam and again allowing the country to become mired in unwinnable wars.

Although Vietnam often takes the central place in research on casualty aversion, there is ample evidence of its influence in other conflicts as well. Choices made during virtually every American military operation since Vietnam have been linked to this phenomenon. Somalia became another of the recurrent examples of the effects of casualty aversion when the loss of eighteen American soldiers precipitated a withdrawal from the country.[9] Moskos goes so far as to say that the conflict created a "Somalia syndrome" to replace the Vietnam syndrome in demonstrating that the American public will not tolerate casualties.[10] Sapolsky and Shapiro echo this sentiment when they say that "one very bad night was enough to start our planning for a withdrawal from Somalia."[11] The collapse of support for the intervention in Somalia is remarkable in these accounts because it came suddenly and after a single incident, rather than following years of smaller declines in support, as in Vietnam.

Mueller's work has remained central to sustaining the view that civilians are highly sensitive to wounded and dead soldiers. Most recently, he has argued that the war in Iraq showed the same basic process of support falling as casualties increase, though in this instance he sees the process occurring even more rapidly than in the past.[12] Luttwak writes of the emergence of post-heroic warfare in an era when "military force collides with the general refusal of the American public to sanction interventions in place after place without end."[13] As he sees it, behind the public reluctance to fight is the desire to avoid loss that has become stronger with the shrinking size of families. The implication is that victory requires the ability to withstand casualties and to persevere undaunted.

Some have argued that the evidence for casualty aversion is less straightforward. Burk finds evidence of rapid fluctuations in support for the intervention in Lebanon and a counterintuitive increase in Americans' willingness to fight after the bombing of the Marine barracks. He likewise identifies a drop in public support for fighting in Somalia taking place before the battle that left eighteen soldiers dead, then being erroneously cited as an effect of the battle. This leads him to conclude that "patterns of public support for peacekeeping missions reveal no irrational or kneejerk reactions based on a putative unwillingness to tolerate casualties."[14] Eric Larson disputes the evidence that the American public has a low tolerance for casualties and argues that "it is more accurate to say that the public hopes for low-to-no casualty operations but fears a very different outcome."[15] Here casualty sensitivity simply appears as a preference and not as a decisive factor influencing American foreign policy. Charles Hyde goes even further by characterizing casualty aversion as a myth and saying that the members of the public make careful evaluations about conflicts that are guided by rational decision procedures and cues from leaders.[16]

Much of the recent work on casualty aversion has sought to show that it is a more complex process than one of casualties producing a clear decline in support for war. Some have even given renewed attention to the early works on casualty aversion to show that the formative studies of this phenomenon made more complex predictions than this type of straightforward drop.[17] Other studies have identified additional intervening variables. Kriner and Shen contend that the strength of casualty sensitivity varies depending on where the casualties come from, with those communities that have suffered disproportionately in the past being less willing to support new military ventures.[18] This has regional and socioeconomic manifestations that reflect disparities in where members of the military come from. This builds on work from Gartner, Segura, and Wilkening that shows similar differences in support for the Vietnam War depending on the extent of suffering communities endured in the fighting.[19]

Delany finds that multiple factors influenced the decision to withdraw American forces from Somalia, and that the growing number of casualties was only the final cause in building a consensus that the war, and the project of nation-building more broadly, was not in the country's interests.[20] Gelpi, Feaver, and Reifler argue that casualty sensitivity does have some influence on support for wars, but that it is only one of a cluster of factors that voters take into account and that it is outweighed by the perceived likelihood of success and attitudes toward a war's justifiability. We will return to these other causal mechanisms over the following chapters, paying special attention in chapter 5 to anticipations of success and failure. For now, the critical point is that these authors find some decline in support for war as casualties mount, but that it is rarely as sudden or as clearly linked to military losses as the conventional wisdom about the casualty sensitive American public would have it.

Some of those who suffer the most during wars, such as those who are injured or whose family members and friends are injured or killed, may not feel casualty aversion. Boettcher and Cobb argue that support for the war in Iraq remained high enough to sustain military operations and even to secure George Bush's re-election because of the influence of the sunk costs trap.[21] Far from triggering a withdrawal, the losses led many policymakers and civilians to call for a sustained investment in Iraq that would redeem the losses and ensure that they were not in vain. The military's response to early defeats was, in General Petraeus' words, to "go all in" in an effort to reclaim the losses, a decision that prompted the troop surge of 2007.[22] This research reveals that the interplay of casualties and support for war is far from certain and that much disagreement remains about the mechanisms driving shifts in approval for fighting.

Casualty Aversion and Restraint

In Korea, Vietnam, Lebanon, Somalia, and Iraq, most of the evidence supporting low casualty tolerance emerged *during* the fighting. The typical approach to studying casualty sensitivity in these cases is by comparing casualty figures to polling data, sometimes while situating these figures in the context of events taking place during the war that may offer insight into intervening variables. For example, studies may show a steady decline in enthusiasm for fighting in Vietnam, then a sharp drop following the perceived American loss of the Tet Offensive, or they show that support for the intervention in Somalia fell gradually and that the Battle of Mogadishu did not actually disrupt the existing trend. Whatever the cases involved, this method of searching for the effects of casualties as they occur is in line with Mueller's discovery of a steady decline in support for war as more American soldiers were injured or killed. However, this is only one possible way of interpreting the evidence.

In addition to casualties causing drops in public support during a conflict, we may also see casualty aversion in relation to *anticipated* losses. Here casualties are a constraint on when and how wars can be initiated, rather than a force compelling the end of conflicts already in progress. Many studies of American foreign policy throughout the 1990s link military restraint and an unwillingness to effectively prosecute operations with a fear of the public backlash that might occur.[23] The American interventions in the Balkans stand out as some of the strongest evidence of anticipated casualties restraining military operations. Ignatieff says of the intervention in Kosovo that "[i]t was fought without ground troops, in the hope and expectation that there would be no casualties at all." In many ways this is an even more powerful manifestation of casualty aversion than drops in support that occur during a war, since it may seriously constrain when and how military force is used rather than just limiting the duration and extent of operations.

The effects of casualty aversion should be seen not only in the public responses to losses that actually occur, but also in the pressures of avoiding those casualties in the first place. The problem is that the effects of casualty aversion on future actions is difficult to gauge because it requires exploring counterfactual scenarios. As Smith points out, "it is . . . easy to claim that a decision not to make a commitment is based on fear of casualties and difficult for a government to deny such a charge, since it has to prove a negative."[24] It remains unclear whether reservations about intervening in Kosovo or starting other prospective American military operations were informed by fears of military losses or by other considerations. It is especially problematic to say when, if ever, the fear of losing soldiers has compelled politicians to avoid fighting. With no casualty figures to link with polling data in these instances, it is impossible to tell whether shifts in support for a potential war are driven by the anticipated casualties or by some other factor. This challenge informs our methodological approach, which we describe in detail later in the chapter, of employing survey experiments about hypothetical conflicts. These experiments allow us to gain deeper insight into *prospective* conflicts, which cannot be judged with polling data because of their hypothetical character.

The arguments mobilized to support drone operations reveals that these machines are often justified based on their ability to reduce the risk to American military personnel. Strawser argues that armed forces have a moral obligation to protect their personnel to the greatest extent possible.[25] It would, he thinks, be irresponsible and morally questionable for them to deliberately expose soldiers to the hazards of the battlefield when there are technologies for improving their security. Extrapolating from this, it is possible that the availability of drones could promote a sense that it is unfair to risk soldiers' lives unnecessarily. Sapolsky and Shapiro reach a similar conclusion when speaking about new military technologies more generally.[26] They find that each time the United States reduces the number of civilian and military casualties incurred by an operation, it also raises expectations for future operations. They envision a kind of ratchet in which policymakers and members of the military must continually strive to either improve on past performance or face a public backlash for not meeting expectations. This suggests that the availability of drones and their recent use in relatively low-cost operations (at least to American military personnel) may intensify preferences for drones and other precision weapons compared to employing manned aircraft or ground forces.

In much of the literature, casualty aversion is characterized as an undesirable phenomenon that has high political and strategic costs for the United States. Many of the commentators who find evidence that casualty aversion is overstated or that it does not exist explain the persistent belief in it as a misperception by elites. Gelpi, Reifler, and Feaver, who find mixed support for casualty aversion, contend that "the view that the public is casualty phobic is widely entrenched among policymakers and the elite."[27] Building on his contention that the phenomenon is largely illusory, Hyde says that there is "strong evidence that policymakers and senior military leaders believe the American public is casualty averse and will not tolerate deaths except when vital interests are at stake."[28] He goes on to say that "[b]y attributing casualty aversion to the public, civilian and military elites have masked their own aversion to casualties and threatened our status as a superpower."[29] Work on American military strategy throughout the 1980s and 1990s reveals widespread concern over the prospects of eroding civilian support and, in particular, of another Vietnam. Policymakers' reticence to lose soldiers is also evident in their rhetoric and their decisions about whether to support wars.[30] Kull and Destler report that the desire to avoid losing soldiers in combat was a theme of their interviews with senior government officials.[31]

The perception that the American public has limited tolerance for losing soldiers extends beyond American officials. Gentry finds evidence of figures as diverse as Yugoslav, Syrian, and Iraqi leaders, Iranian and Chinese generals, and terrorists predicting that they could defeat the United States by inflicting light casualties and undermining public support for fighting.[32] Osama bin Laden famously claimed that the withdrawal from Somalia demonstrated the country's unwillingness to pay the costs of war.[33] These perceptions even play a role in negotiations, as in the case of a Chinese official refusing to back down during a 1996 confrontation over Taiwan, saying, "We've watched you in Somalia, Haiti, and Bosnia, you don't have the will."[34] Sapolsky and Shapiro contend that "we have grown ever more sensitive about casualties . . . and we seek to avoid them. This limits our ability to exercise the tremendous power we possess and makes us susceptible to pressures others can ignore."[35] With these points in mind, some conclude that perceived casualty aversion on the part of American and foreign officials will limit foreign policy choices, slow military responses, and undermine the country's ability to exercise deterrence.[36] Thus, according to some, casualty aversion is a potentially dangerous phenomenon that reduces security and should be overcome.

Critics of drone warfare tend to take a much different perspective on casualty aversion, seeing it as a beneficial constraint on military operations that must be sustained in the interest of preventing unnecessary or aggressive wars. We already mentioned Singer's concern that using drones to avoid costly military operations could trivialize conflicts. He also says that when the costs of war diminish, leaders gain a dangerous freedom of action that could lead them to initiate conflicts at will: "[a] leader needn't carry out the kind of consensus building that is normally needed before a war, and doesn't even need to unite the country behind the effort."[37]

Others raise similar concerns, making this one of the most powerful arguments against drone warfare. Kaag and Kreps argue that public opposition to losing soldiers is one of the central mechanisms through which citizens force the government to be accountable to its wishes. They contend that the pressure to avoid costly wars is responsible for the growing use of drones, which will ultimately succeed in limiting public engagement in use of force decisions and degrading accountability. They say that "methods of warfare that leverage technology in order to insulate citizen-soldiers from harm" and that this "creates the possibility that leaders will no longer, in a prudential sense, have to obtain popular permission to go to war."[38] Linda Johansson argues that pilot invulnerability "might have an impact on domestic opinion, and, in turn, lower the threshold of entering and sustaining a war."

Frank Sauer and Niklas Schörnig contend that drones are uniquely attractive to democracies because they allow elected officials to wage wars without the potential adverse consequences that would come from losing soldiers in battle. Thus, as with Kaag and Kreps, they see the invulnerability of drone operators as a threat to accountability and the loss of one of the barriers against initiating wars. Finally, in another variation of this argument, Daniel Brunstetter and Megan Braun conclude that drones make major wars less likely by preventing cycles of escalation between states, but that they will make small wars much easier by lowering the costs of intervention. This suggests that it will be especially important to test the effects of drones on approval for fighting by looking at small military interventions that may only be tenable when the anticipated casualties are light or nonexistent.[39]

Both supporters and critics of combat drones, then, use the fact that these weapons produce no military casualties to bolster their conclusions, albeit with much different goals in mind. The conjecture that drones could make it easier to fight by circumventing casualty sensitivity is plausible, but at present there is no strong empirical evidence supporting it. Those who advance this claim rely primarily on studies of casualty aversion produced before drones were being actively used in combat or that fail to distinguish drones from other attack types. With so many of the moral concerns relating to drones resting on this empirical prediction, it is imperative to gain better understanding of whether drones actually have any effect on perceptions of conflict. Our purpose differs from these other commentators; we seek to determine if it is true that the pilot invulnerability enabled by drones is one of the reasons that individuals, on average, prefer this means of attack over others.

Attack Type and Support for the Use of Force: Survey Experiments

The discussion to this point indicates that concern about military casualties often is a key influence on individuals' attitudes toward the use of military force. All things equal, attacks conducted by drones should receive more support than attacks that expose military personnel to the risk of physical harm, an expectation supported by the conjoint experiment discussed in the previous chapter. To assess this idea, we designed a vignette-based survey experiment that manipulated the type of attack. All respondents read a mock news story that began with the statement that al Qaeda had established training camps in the country of Yemen and planned to use these to attack American targets:

> Terrorists connected to al Qaeda have established bases and training camps in the country of Yemen in the Middle East. Political turmoil has prevented the government of Yemen from acting against the terrorists. Recently the Yemen branch of al Qaeda attempted to bomb an American airliner and to mail bombs to the United States. American intelligence agencies have identified the location of the al Qaeda bases in Yemen.

This introduction was followed with information about how the United States planned to attack these bases. Respondents were randomly assigned to read a paragraph that described the attack as being carried out by drones, by aircraft with onboard crews, or by ground troops:

> *Drone Strike:* The United States plans to launch attacks on these bases with missiles fired from unmanned drone aircraft to kill al Qaeda leaders and militants. The use of unmanned drones means that no American military personnel would be placed at risk.

> *Air Strike:* The United States plans to launch attacks on these bases with bombs dropped from airplanes to kill al Qaeda leaders and militants. The militants are believed to lack weapons capable of attacking these airplanes.

> *Ground Troops:* The United States plans to strike the bases with American paratroopers. These American troops would attack the militants and their leaders located in the bases.

After reading this news story, respondents were asked to indicate how much they supported the attack and to answer some additional questions detailed below. This chapter's second expectation leads us to expect that the average level of support among respondents who read about an attack by drones will be higher than among those who read about an assault with ground forces. Average support among those who read about air strikes should fall somewhere between these two extremes.

We followed previous experiments on support for the use of force in choosing Yemen as the location for this scenario, for two reasons.[40] First, the scenario is a realistic one; al Qaeda has, in fact, used Yemen as a base from which to attack American targets, such as the suicide bombing of the USS *Cole* in 2000 and attempts to bring down airliners bound for the United States in 2010 and 2011. As we discussed in chapter 1, Yemen has also been the site of many American targeted killing operations involving drones, special operations forces, and precision-guided munitions such as cruise missiles and missiles launched from manned aircraft. Respondents are therefore apt to perceive Yemen as a realistic setting for US military interventions and may even recall some of these previous missions.

Second, few Americans are familiar with Yemen, and the country did not receive a great deal of news coverage while we conducted this experiment. Polls of Americans' knowledge of international affairs show that most people have little awareness of events in the country.[41] Over the course of our research, the fighting against al Qaeda in Iraq, then Islamic State in Syria and Iraq, received far more attention than drone operations in other countries. This means that most respondents reading our experimental conditions should base their responses on its content rather than on information they have independently acquired by reading or watching news media (see the chapter appendix for the full texts of all the news story conditions). Even more importantly, there was a risk that fictional scenarios involving more familiar conflict areas or rivals, such as Iran, Iraq, Syria, Afghanistan, and Pakistan, would produce bias because of the greater public awareness of them and the stronger negative connotations. Yemen is therefore a good case in which we can create plausible fictional conflicts without much risk of respondents' background knowledge of the region interfering with their decisions.

The experiment manipulates how the United States plans to carry out attacks on the militant bases, which we call the attack type. While all respondents' news stories contained identical background information on the situation in Yemen, the attack type is systematically varied. Respondents were randomly assigned to news stories that described the attack as being carried out by "unmanned drone aircraft," "bombs dropped from airplanes," or "American paratroopers." The first two of these conditions included information about the likelihood of military casualties. The drone strike condition stated that "[t]he use of unmanned drones means that no American military personnel would be placed at risk." The air strike condition included the information that "[t]he militants are believed to lack weapons capable of attacking these airplanes."

One concern is that including this information about the chance that Americans service members could be harmed might draw the attention of some respondents to the issue of military casualties, reducing their willingness to support the attack. Previous research on framing, especially work grounded in prospect theory, has found that emphasizing losses can prompt a risk-averse style of thinking.[42] When it comes to the decision to initiate war, this work has found that presenting the adverse effects of fighting, which include not only the expected casualties but also other negative evaluations like a low likelihood of success or high civilian casualty estimates, can reduce support for war by encouraging members of the public to focus on the possible bad consequences of fighting over the benefits that it may offer. This suggests that explicitly mentioning expected military casualties could cause a decline in support for military operations.[43] Studies of framing that deal with drone strikes and autonomous weapons have likewise shown that framing plays an important role in activating risk aversion. When questions mention the human costs of launching attacks or normative transgressions, respondents become less supportive of launching attacks.[44]

We decided to include information about expected casualties in the treatments for three reasons. First, doing this ensures that respondents understood this element of each attack type. Our worry was that not including these details would lead respondents to pay little attention to the *type* of attack that was planned and instead focus on the more general issue of whether it was a good idea to use military force in this situation. This is particularly important when it comes to evaluating support for drones. These weapons platforms are relatively new, having only been used to launch attacks for around ten years at the time of the study. Without some mention of the potential risk of military casualties, it may be difficult for some respondents with little knowledge of military affairs to make informed risk evaluations when drones are involved.

Second, the conjoint experiment discussed in chapter 2 shows that not including information about the risk of harm that military personnel face does not alter the findings we report below. That experiment also varies the attack type, but the conditions include no mention of the chance of military casualties. The results of that experiment are consistent with those we discuss in this chapter, lending credence to the idea that including this information did not decisively influence our findings.

Third, if explicit mention that drone strikes will not result in military casualties and that air strikes might result in casualties were to have any influence on respondents' risk evaluations, then the existing research indicates that the effect would be to increase the divide between these two treatments.[45] Explicitly mentioning the possibility of casualties when it comes to air strikes is apt to trigger a risk-averse stance, while ruling out that possibility when it comes to drone strikes should trigger a risk-acceptance stance. If these different attitudes are activated, then the effect would be to increase the divide between the drone and air strike treatments. As we will discuss later in the chapter, the separation between the drone and air strike treatments is fairly small, indicating that any influence of risk framing on evaluations was likewise weak or nonexistent.

After reading the news story, all respondents answered the same four questions.[46] The first asked them to rate the attack plan on a seven-point scale ranging from "definitely do not carry out attack" to "definitely carry out attack." Responses to this question are used to assess the chapter's first expectation. They were then asked to estimate how likely it was that the attack would result in military casualties, achieve its military objectives, and kill civilians. We included these questions to evaluate the chapter's second expectation, which holds that respondents assigned to the drone strike condition would predict that there would be fewer military casualties than those in the ground troops condition. As discussed in chapter 1, some proponents of drones suggest that in addition to reducing military casualties, their technical capabilities, such as the ability to integrate real-time intelligence into the timing and targets of attacks, make them both better at achieving military objectives and less likely to cause civilian harm. The last two questions assess whether respondents assigned to read about drone strikes were more likely to think the attack would succeed or avoid civilian casualties than those who read about an air strike or the use of ground troops.

We conducted this experiment with two respondent pools about one month apart.[47] The first was recruited by the survey research firm YouGov and is a sample of 1,000 adults that is representative of the population of the United States.[48] The second was a sample of 1,202 adults recruited via Amazon's Mechanical Turk online labor market. Duplicating the same experiment with these two respondent pools allows comparison of how nationally representative samples, such as our YouGov respondents, and respondents in convenience samples respond to the same manipulations. Respondents in convenience samples such as Mechanical Turk may respond to experimental treatments in different ways than do respondents in more representative samples. In later chapters, we rely on Mechanical Turk samples to assess hypotheses about the effects of drone warfare on attitudes. The identical experiments analyzed in this chapter produce similar findings across the two different samples. This suggests that the findings of our subsequent work that uses convenience samples alone is likely to reflect mechanisms that have a strong potential to be generalized to the public at large.

We preregistered our plan for collecting and analyzing data prior to fielding the experiments.[49] Preregistration strengthens confidence that our findings, and the similarities across the representative and convenience samples, are not the product of post hoc decisions about which measures to include or how to conduct the analysis. Our pre-analysis plan identifies the dependent variable as responses to the question asking about support for the attack described in the news story, and ordered logistic regression as the estimator. We follow the convention of using weights for the data from the representative sample but not the convenience sample.[50] The key independent variables are dichotomous variables identifying if respondents were randomly assigned to the air strike and ground troop conditions; assignment to drone strike treatment is the excluded category.

Our expectation is that the coefficients on the ground troop variables will be negative. We also include standard covariates for both samples: gender, party identification, age, education, and identification as a member of a minority group. The chapter appendix compares the political and demographic characteristics of the representative and convenience samples to those of the American National Election Survey (ANES), which has rigorous procedures for selecting a random sample of adults. Our YouGov sample is similar to the ANES sample, while the Mechanical Turk respondents for this experiment are, on average, younger and more likely to have a college degree, to identify as agnostic or atheist, to be Caucasian, and to be Democrats. These differences are typical for such samples.[51] For this convenience sample, we run an additional model that includes five additional covariates: age squared, income, marital status, political ideology, and identification as atheist or agnostic. Previous research finds that including these covariates produces estimates of causal effects in Mechanical Turk samples that are qualitatively similar to those produced when identical experiments are conducted with representative samples.[52] This allows us to further identify the degree to which our substantive findings differ across these two sampling frames.

Figure 3.1 plots the percentage of support offered by respondents in our YouGov sample in each of the three conditions.[53] Two facts stand out. First, large fractions of respondents support the use of force across the three attack types. This is consistent with previous work that has found that Americans are particularly willing to support military action to counter terrorism directed at the United States.[54] Furthermore, the experiment was conducted at a time when the actual threat of terrorist attacks against the United States and its interests was quite high. Over the previous year, Islamic State–inspired terrorists had launched attacks in the United States, Belgium, and France, all of which received extensive press coverage.[55] Based on this we can surmise that casualty sensitivity may be a barrier to initiating wars, but that perceived national security imperatives can still be strong enough to push the public to support war despite the possibility of losing soldiers. Even after more than a decade of intractable fighting in Iraq and Afghanistan, leaders could plausibly rally sufficient backing to send American forces into another conflict. It is essential to avoid overstating the extent of casualty aversion or underestimating the power of perceived national security priorities when it comes to mobilizing support for war.

Second, differences across the conditions are modest; of the respondents who read about an attack by ground troops, a slightly smaller percentage supported attacking, and a slightly larger percentage opposed attacking, compared to those who read about attacks by drones or aircraft. This is consistent with our first expectation that attack type influences support for the use of force, but suggests that the substantive effect of variation in attack type may be smaller than indicated by those who fear that

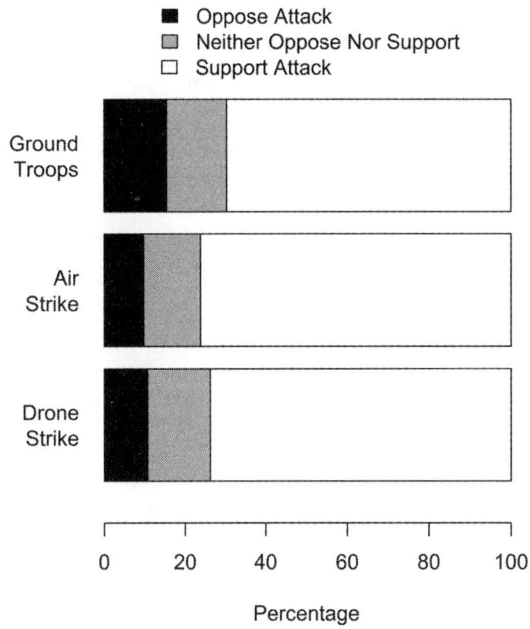

Fig. 3.1. Support for use of force by attack type

using drones will substantially increase public support for military action. The differences across attack types has several important implications. It is further evidence that casualty aversion is not an overwhelming impediment to initiating wars. Fear of loss appears to shift levels of support, but only by a modest amount. It seems doubtful that such small changes will have a decisive impact on whether the country goes to war, especially when leaders have a degree of autonomy from public opinion and may initiate wars even when they lack majority backing. It is possible that casualty aversion could have a stronger limiting influence in other countries, but in the case of the United States there is little empirical basis for the concern that drones will seriously alter the public's willingness to fight terrorism with military force. This sheds light on the moral debate over drone warfare by challenging the widespread assumption that drones make it easier to fight.

The similarities between treatments involving drones or inhabited aircraft give us greater insight into how the American public perceives the risks associated with war. Drones may in principle mark a radical step beyond inhabited aircraft because they make pilots invulnerable, but from the public's standpoint, all air warfare seems to be relatively safe.

This is likely a product of decades of American air superiority. US fighters and bombers have dominated the skies against enemies with limited anti-aircraft resources. Only a few pilots have been shot down in recent conflicts, with most aerial casualties coming from friendly fire and accidents. Moreover, survivability is high for combat aircraft. Since the end of the Cold War, many pilots who have been shot down by enemy fire have managed to escape and were later rescued. As long as the US military is preoccupied with fighting insurgents and terrorists, the comparative risks associated with drones and inhabited aircraft will continue to be low. These will only change substantially if the United States begins operations against conventional enemies, though even in this case the public's background assumptions about risk may persist until the country begins losing pilots in action.

Returning to the debate over the morality of drone warfare, the close relationship between attitudes toward both types of aircraft is evidence that the real challenge related to the disappearance of casualty aversion is not coming from drones. Rather, it is a result of asymmetric balances of power between opposing belligerents. So long as inhabited aircraft are virtually invulnerable from enemy fire, they present roughly the same advantages as drones. The possibility of harming pilots onboard aircraft does not matter much when the pilots are rarely harmed in practice. The ability to circumvent feelings of casualty aversion are therefore not fixed and unchanging. They depend heavily on enemy military capacities. Expanding on this, other vehicles that offer similar asymmetric advantages or heightened survivability for crews could have the same capacity to moderate perceived risks as drones themselves.

In terms of bypassing casualty sensitivity, the unique advantages drones offer will only be strongly felt when comparable inhabited vehicles are threatened. As discussed in chapter 1, current generation drones will probably have limited utility in these conflicts because they are more susceptible to enemy aircraft and anti-aircraft systems than planes with onboard crews, which are faster and have more sophisticated countermeasures. Thus, the advantages drones confer when it comes to generating a consensus for war will probably be most valuable when leaders are less likely to consider these to be the optimal means of attack.

Table 3.2 takes up the issue of casualty sensitivity more systematically. It reports the results of four ordered logistic regression models, using support for the attack as the dependent variable. Models 1 and 2 use data from the representative YouGov sample; models 3 and 4 use data from the Mechanical Turk convenience sample. In model 1, the only independent variables included are dichotomous measures of assignment to the air strike or ground troops conditions. Model 2 adds the covariates outlined in our pre-analysis plan. Models 3 and 4 report similar analyses for the convenience sample; model 3 uses the same covariates as model 2, while model 4 adds the covariates that previous research suggests will produce results most similar to those in the representative sample. Across the four models, coefficients on the air strike variable are not significant, indicating that, on average, respondents in our two samples did not systematically prefer drone strikes to attacks from manned aircraft. The coefficients on the ground troops variable are negative and statistically significant in all four of the models. Our respondents, then, do prefer that attacks be carried out by drones rather than by ground troops.

Turning briefly to the control variables, we see results that are broadly consistent with previous work on support for the use of force. Individuals whose party identification is Republican express more support for attacks in models 2 and 3. This variable is not significant in model 4, but the measure for political ideology it includes is positive and significant, indicating that more politically conservative respondents are willing to engage in the use of force. Individuals who are older also express more support for military action, as indicated by the age variable in models 2 and 3 and the square of this variable in model 4. Women are less likely to express support for the attack, although this variable is only significant in the convenience sample.

The final row of table 3.2 reports the odds ratios for the ground troops variable in models 2, 3, and 4. These indicate the change in the likelihood that a respondent would select the highest level of the dependent variable ("definitely attack") compared to all of the six lower levels of support for military action, holding all other covariates at their means. This sort of comparison is a meaningful one, as we saw in figure 3.1 that pluralities of respondents in each of the conditions endorsed launching the attack that they read about. For model 2, for example, individuals who read about the use of ground troops were 28 percent less likely to offer the fullest degree of support for the attack than those who read about the drone strike. The corresponding odds ratios for models 3 and 4 are of similar magnitude.

We saw in figure 3.1 that there are small but discernible differences in the attitudes of respondents based on the content of the news story they read, with those who read about an attack by drones being more willing to support the use of force. The statistical analysis sheds additional light on this relationship. It tells us that the differences we observe in figure 3.1 between respondents assigned to the drone and ground troop treatments are statistically significant (albeit at the $p < .10$ level in model 1), but that respondents who read about drone strikes and air strikes offered similar levels of support for the planned attack. This relationship holds when we include control variables such as party identification, meaning we can be more confident that it is the attack type and not some difference in indi-

TABLE 3.2. Ordered Logistic Regression Models of Attack Type and Support for the Use of Force

	Representative Sample		Convenience Sample	
Model	(1)	(2)	(3)	(4)
Air strike	.03	.10	–.19	–.19
	(.14)	(.15)	(.13)	(.13)
Ground troops	–.24*	–.33**	–.46***	–.42***
	(.14)	(.15)	(.13)	(.13)
Gender		–.10	–.18*	–.19*
		(.12)	(.10)	(.11)
Party identification		.11***	.24***	.07
		(.03)	(.03)	(.05)
Age		.03***	.02***	–.05
		(.004)	(.005)	(.03)
Education		–.09**	.06	.03
		(.04)	(.04)	(.04)
Minority		–.21	.24*	.23*
		(.13)	(.13)	(.14)
Age squared				.001**
				(.0004)
Income				.09***
				(.02)
Married				–.05
				(.12)
Atheist or agnostic				–.04
				(.12)
Ideology				.39***
				(.08)
Observations	994	947	1,202	1,197
Log likelihood	–1,603	–1,471	–2,083	–2,047
Ground troops odds ratio	.79	.72	.63	.65

Note: Standard errors in parentheses; * $p < .1$; ** $p < .05$; *** $p < .01$.

vidual characteristics that determines different levels of support for the attack across treatments.

We also see that the relationship is similar in both the representative YouGov sample and the convenience Mechanical Turk sample. This will prove important in subsequent chapters, where we rely on additional convenience samples to assess our empirical expectations. The fact that respondents in both respondent pools respond in similar ways suggests that the findings of our later chapters have the potential to generalize from our online samples to more representative groups. These experiments indicate that drones have an incremental effect on attitudes and that, as we saw in chapter 2, other factors continue to exert powerful influences over support for the use of force. Thus, while we argue that the risk of drones circumventing casualty aversion has been overstated, we do think that it is important to treat the availability of drones as being one consideration in the overall explanation of what directs public opinion when the country is contemplating war.

To this point, our analysis is consistent with the idea that individuals are a bit more willing to express support for an attack carried out by drones than one carried out by ground troops. But this sheds less light on *why* people favor drones. In chapter 1 we suggested three reasons for this relationship. The first is that drones eliminate the risk of military casualties, and a great deal of research concludes that avoiding such harm is an important consideration when Americans assess the utility of military force. A second is that, given drones' technological capabilities to loiter for long periods and to collect intelligence on targets in real time, individuals might favor drones because they are seen as a highly selective type of military force that is more likely to achieve battlefield objectives. A third, and related, reason is that these capabilities may reduce the chance that drones, compared to other attack types, will cause civilian casualties.

We assessed these arguments by asking respondents how likely they thought that the attack described in the news story they read was to result in military casualties, a successful attack, and civilian casualties. Figures 3.2 through 3.4 plot responses to these questions from respondents from our representative sample. In figure 3.2, we see that respondents were considerably more likely to expect military casualties if they read about an attack by ground troops than by aircraft or drones.[56] This is consistent with the proposition that avoiding such harm influenced the degree to which they supported the attacks. However, no such relationship exists for mission

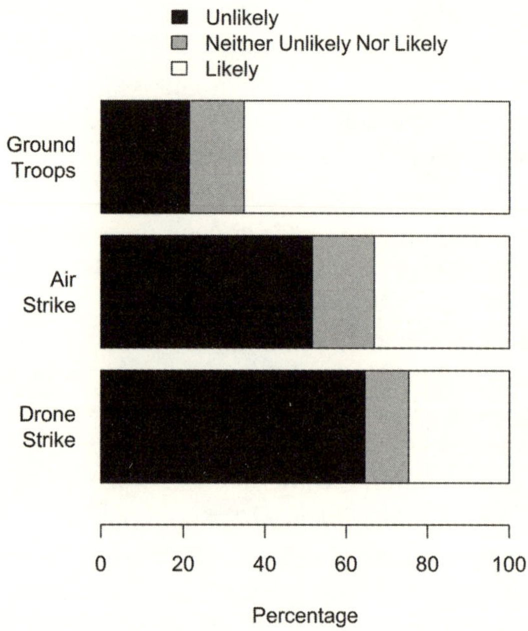

Fig. 3.2. Estimates of military casualties by attack type

success (figure 3.3)[57] or civilian casualties (figure 3.4),[58] indicating that the type of attack exerted little influence on expectations regarding these outcomes. The absence of a relationship between perceptions of mission success and civilian casualties reinforces this point. The fact that these two factors do not vary systematically across treatments, while the expectation of military casualties does, is consistent with our expectation that the latter is an important reason why respondents were more supportive of attacks with drones than with ground troops.

These finding suggest that military casualties matter for respondents' support for the attack. These experiments, however, are not definitive on this point. The reason for this is that respondents in the drone strike and air strike conditions were provided with information suggesting that military personnel would face low risks of harm if the attacks took place, so the relationship depicted in figure 3.2 could simply reflect this feature of the experiment.[59] However, we saw in the conjoint survey experiment in chapter 2 that similar patterns hold when information about military casualties is not included, indicating that our findings in this chapter are unlikely to be a result of this specific design choice.

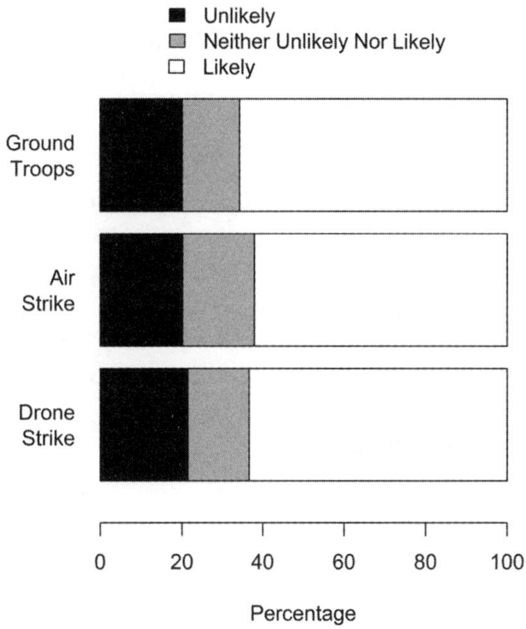

Fig. 3.3. Estimates of civilian casualties by attack type

Conclusion

It is important to reiterate here that our experiments in this chapter and throughout the book focus on how casualty aversion influences the public's support for *new* military ventures. The experiments respondents read involve the initiation of wars or the expansion of ongoing conflicts into new areas. Initial support for war may not always be a good indicator of whether a war will continue to be popular over the long term. As Mueller points out, the wars in Korea, Vietnam, and Iraq had considerable support early on and yet became unpopular once hostilities began.[60] The level of support for a new intervention could therefore drop substantially once the costs become real, especially if these deviate from what elites have promised. Burk even says that the core problem of casualty aversion is not building the support to start an intervention but rather "The worry is that public support for missions, which seems sufficient, will quickly evaporate when faced with American casualties."[61] However, this is likely an overstatement. As we saw earlier, many commentators link casualty aversion to a loss of American deterrent power (which is premised on inhibiting the initial use

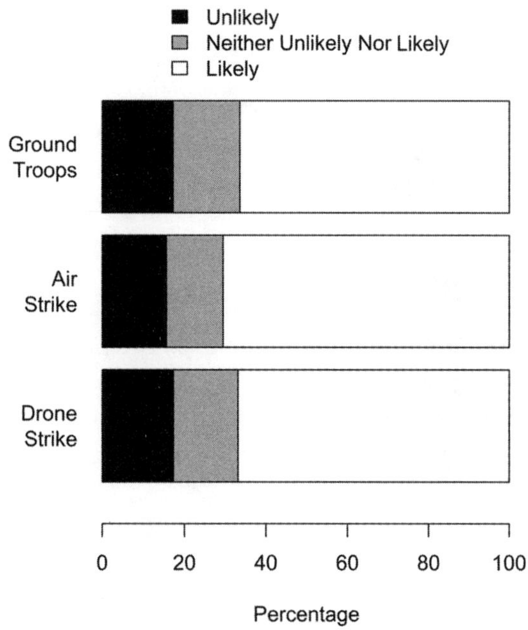

Fig. 3.4. Estimates of success by attack type

of force that we test here) and credit this feeling with producing a spate of low-intensity military operations since the Vietnam War, as well as causing a general reticence to fight when American interests in doing so were weak.

Of course, there is an even more important reason for looking at the beginning of conflicts rather than at changes in support for ongoing wars. Thus far our only experience with American drone strikes has come in the context of a War on Terror that is difficult to characterize as a single coherent war, at least not in the same sense that Vietnam or Somalia were self-contained wars. As discussed in chapter 1, drone strikes have been characterized not by sustained operations but by sporadic attacks taking place around the world and against an ever-changing array of targets. In this context, judgments about the permissibility of fighting must often be made on a case-by-case basis. They continually arise as attacks expand geographically and as new targets are selected. That is to say, during the War on Terror, the American public continually faces new conflict scenarios akin to those we describe in the experiments even though these strikes are ostensibly part of an ongoing war.

Recall that one of the central concerns among those who consider casualty aversion to be an unfortunate limitation on American foreign policy was that it (or at least perceptions of it) prevents the United States from coercing rivals. Whether the United States actually engages in coercive deterrence will depend heavily on whether elites sense that they can risk committing military forces and still be assured of public support. The enduring gap between elite perceptions of civilian casualty sensitivity and the public's willingness to sustain some losses is therefore an important consideration when anticipating what effects public opinion will have on policymakers' choices. Our experiments do not gauge elite attitudes toward drones, yet this issue of elite perception and its relationship to casualty aversion shows that any effects of public opinion will continue to be influenced by and mediated by elites. The consequences of casualty sensitivity for the American capacity for deterrence are therefore apt to vary considerably depending on the views of American officials and of those on the receiving end of coercion.

It is also important to be aware that the effects of casualty sensitivity may vary over time and continue to change as the public gains familiarity with drones. Levy argues that sensitivity to losses only empowers civilians in the short term, and that over time the restrictive force of public opinion dissipates as the military seeks ways of circumventing it. In particular, military elites may attempt to speed up the decision-making process to outpace public opinion formation, create contingency plans that weaken civilian oversight, or increase their influence over the decision-making process. Drones themselves may arguably be one means of re-empowering the military and intelligence services, as they seem to fit closely with the process Levy describes as reducing risks to soldiers with the goal of increasing the military's freedom of action. Drone warfare could therefore be a consequence of casualty aversion and an effort to reassert the military's ability to wage war without being constrained by external pressures.

Although Levy sets out to challenge the dominant understanding of casualty aversion, his analysis supports the general conclusion that critics of drone warfare tend to draw from that literature: that drones threaten to produce a dangerous drop in civilian concern over the costs of military actions. With this in mind, we should not assume that support for drones compared to other types of military force will remain consistent over time, and we should continually revisit this topic to gauge shifts in public opinion. This will be especially important following any major security events that can dramatically reorient national security priorities, as the War on Terror did.[62]

The focus on the effects of military casualties in this chapter is essential for determining whether the casualty aversion thesis is plausible and what effects casualty sensitivity may have on the future of American drone operations. By isolating this factor and testing it apart from those we discussed previously, it is possible to confirm that judgments of prospective military operations are informed by assessments of military casualties, but that the effects of casualty aversion on overall levels of support for war is small. Over the following chapters we will introduce other factors that produce a more comprehensive account of how it operates and what effects it will have on the future of American military operations.

CHAPTER 4

Drones and Policy Objectives

American combat drones have been used extensively to target militants and terrorists in countries such as Pakistan and Somalia. Less frequently noticed is that they have also been employed for other purposes. As we discussed in the first chapter, the United States used drones during the civil war in Libya in 2011. At the time, there were no terrorist organizations there that seriously threatened the United States. Instead, the goal of the intervention was to assist the rebels seeking to overthrow Muammar Gaddafi. American officials believed that without decisive action on their part, the Gaddafi regime was likely to massacre civilians, and they cooperated with other countries to weaken Gaddafi's forces and to impose a no-fly zone.[1] Drone strikes in the war in Afghanistan have been used to target armed groups, such as the Taliban, that threaten American ground troops in the country.[2] But an additional goal of such strikes has been to bolster the Afghan government in its struggle with the Taliban. In Libya, then, drones were used to bring about regime change and prevent a humanitarian disaster; in Afghanistan, they were used to prevent the violent overthrow of an ally. This indicates that drones can play an important role in a broad range of missions beyond the targeted killings with which they are most commonly associated.

Combat drones have not been employed in conventional conflicts with states and are currently less suited for this role because of their vulnerability to enemy aircraft and air defense systems.[3] Nevertheless, the United States is investing heavily in developing drones that could penetrate opponents' air defense and launch attacks against ground targets or ships.[4] While drones have been used most extensively to counter terrorist groups, there have already been cases where they have been employed to achieve other political objectives, and in the near future the range of operations that drones could undertake may expand rapidly as their technological capabilities are developed further. In this chapter, we investigate if and how the type of mission, or principal policy objective, influences public support for military action when drones are employed.

We also consider how policy objectives shape Americans' willingness to bear losses. Casualty aversion is generally described as constraining uses of force and forming a barrier against the initiation of wars, often without any discussion of whether the type of war in question will influence public support.[5] Those who discuss the possibility that drones may lower inhibitions against initiating wars generally do not account for the many different reasons why wars are waged and how these reasons may affect the public's willingness to fight and sustain casualties. This leads critics of drone warfare to characterize the technology as increasing the incidence of wars or lowering inhibitions against fighting as though the objectives being pursued have no influence on public opinion. This chapter demonstrates that public support for wars is a complex phenomenon, and that other considerations aside from the anticipated numbers of military casualties can influence people's support for fighting. Drones may help to avoid the antiwar sentiment generated by casualty aversion, but this has limited influence on the overall levels of support for fighting when important policy objectives are at stake.

Distinguishing between types of wars is also essential for understanding the political, moral, and legal implications of drone use. All wars are not equal. One war might be seen as less morally and legally justified than another, and the desirability of the political goals being pursued may vary.[6] Critics of drone warfare tend to imply that all efforts should be made to avoid war under all circumstances and that any effect drones have on facilitating combat is undesirable. Conversely, those who support drone warfare tend to emphasize drones' ability to conduct selective attacks while downplaying the possibility that those attacks may lead to more aggressive wars. In both cases there is little attention to the underlying reasons for fighting or how the justification for war may change depending on the circumstances. It is not enough to know whether drones make it easier to initiate wars; we must know whether they make it easier to initiate just or unjust wars in particular.

Wars should usually be avoided, but they are sometimes necessary, and it is therefore best to adopt a more pragmatic attitude when assessing the implications of drone warfare and the effects of casualty sensitivity more broadly. Judgments about whether inhibiting or enabling wars is good or bad must be made on a case-by-case basis with careful attention to the reasons for fighting and the availability of nonviolent alternatives. This shifts the moral and legal discussion of drone use away from the machines themselves and onto the people who will authorize drone strikes. Accepting that attacks are sometimes warranted makes it imperative to determine whether attitudes toward drone operations will be shaped by prudential calculations about when wars are necessary and justified, or by the ease of conducting attacks. If support for drone strikes is influenced by the value and legitimacy of the objective being pursued, then we should expect public opinion to manifest a fairly responsible attitude about drone use. On the other hand, if the public is sensitive to casualties but not to the reasons for fighting, then drones strikes could potentially achieve the same level of support regardless of whether there are good reasons for fighting.

We build on a large body of research that categorizes different types of wars in terms of their principal policy objective (PPO). PPOs can be grouped into four general types: counterterrorism, foreign policy restraint, internal political change, and humanitarian intervention. Theoretically, organizing conflicts in terms of PPOs provides a way of mapping how support for drone strikes is influenced by the kinds of conflicts in which drones could be used. Methodologically, the literature showing how PPOs influence support for wars involving ground forces and air strikes facilitates comparisons between our own research and previous studies exploring other influences on public opinion. We build on the insight that wars that are perceived as being more important for national security tend to receive higher levels of support. We expect that PPOs will affect the willingness to use drones and that support for them will be particularly strong for counterterrorism operations because of the prominence of terrorism as a security threat.

In this chapter, we contribute to existing work regarding PPOs to investigate two expectations, summarized in table 4.1. The first is that the pattern of greater support for drone strikes that we found in chapters 2 and 3 for counterterrorism operations is replicated for other PPOs. As we discuss below, this relationship is not straightforward; there are plausible reasons to think, for example, that pilot invulnerability in particular could make drones especially popular for less intensive military engage-

TABLE 4.1. Principal Policy Objective Expectations

Expectation 4.1: Individuals will express greater support for prospective military missions employing drones compared to other attack types for all principal policy objectives.
Expectation 4.2: Individuals will express greater support for attack types that place military personnel at risk when the principal policy objective represents a more important national interest.

ments. The second conjecture is that the additional support for the use of force that materializes when drones are employed is actually the smallest for counterterrorism operations. In the contemporary era, terrorism is the most direct armed threat to the security of the United States. For this reason, we expect individuals to be more willing to incur costs, such as military casualties, to counter terrorism than they are to address other threats. The survey experiment that makes up the empirical section of the chapter finds strong support for the first conjecture, and less overwhelming but consistent support for the second. In the conclusion, we discuss some of the implications of these findings, including that drones appear to make all types of combat politically easier and thus could, at the margin, enable leaders to engage in more wars of more types in the future.

Drones and Principal Policy Objectives

A central criticism of drones is that the pilot invulnerability and selectivity they allow will lower inhibitions against initiating wars. Most of the commentators who raise this concern do so without considering how different types of PPOs could influence public opinion. Their work implies that there should be a consistent pattern of drones circumventing limits on support for the use of force, and therefore producing greater willingness to wage wars compared to when piloted aircraft or ground forces are used, regardless of the reasons for fighting. If drone warfare indeed renders the population complacent and disengaged, then the PPO may not matter at all, since a disengaged public would be indifferent whatever the circumstances surrounding an operation. Our findings across multiple experiments challenge the assumption that drones produce an acquiescent and uncritical citizenry and lead us to suggest that a more plausible expectation: that the pattern of greater support for drone strikes than other types of force persist across military operations in pursuit of different PPOs. If this is the case, then we would expect to see that although variation in PPOs influences support for war, pilot invulnerability and selectivity will guarantee that drones are consistently preferred over alternative methods of attack that could put soldiers at risk.

Others suggest that the effects of drones on support for military action should vary across PPOs. Brunstetter and Braun are among the few commentators who have considered the effects that drones might have on different types of military operations. They argue that drones will not simply increase the incidence of wars, but rather decrease the likelihood of major wars while increasing the likelihood of small wars. Although their distinction between major and small wars is not identical to the four types of PPOs, their argument does suggest that the characteristics of the conflict will alter how drones influence public opinion. Their reasoning is that drone use will avert large wars by allowing states to engage in small attacks in pursuit of limited objectives and without endangering their military personnel. Such attacks would be less apt to escalate into major conflicts than attacks involving human soldiers because, due to pilot invulnerability, the attacks may not cause any casualties.[7] By contrast, Brunstetter and Braun contend that small wars will become more frequent because policymakers and military elites will be more tempted to use force when they can wage risk-free asymmetric wars. This is a plausible expectation, but the authors do not substantiate it with empirical findings, so more work needs to be done to test their hypothesis.

According to Brunstetter and Braun, the absence of casualty aversion plays an important role in enabling small wars to take place. Public opinion is less likely to inhibit these types of wars when drones are involved than when soldiers may be killed or wounded. This means that casualty aversion could be avoided entirely in small wars, even though it would continue to act as a barrier against initiating more destructive major wars. Although Brunstetter and Braun do not consider how support for drone strikes may interact with differences in PPOs, their argument provides a basis for thinking that the public will be less likely to favor drone strikes when a PPO involves the risk of escalation into a conventional war, as it might in foreign policy restraint and internal political change missions, than when a PPO would typically lead to a small war, as in the case of counterterrorism operations and humanitarian interventions.

Zack Beauchamp and Julian Savulescu agree that pilot invulnerability could make it easier for states armed with drones to wage wars, yet they reach conclusions about the desirability of this that differ from most other commentators engaged in the debate. As they see it, "the bug is in fact a feature" because lowering the threshold for war means that "drones have the potential to significantly improve the practice of humanitarian intervention."[8] It is difficult to cultivate public support for humanitarian interventions because they can have high costs and little or no clear material payoff for the intervening state. Ideally, humanitarian interventions should not be conducted to benefit the intervening country, as this might compromise the moral grounds for intervention and encourage aggression under the pretext of humanitarianism. But as Beauchamp and Savulescu correctly point out, such benevolent missions are unlikely to be undertaken unless they are relatively cheap, and they will only be cheap if they can be fought using military hardware that can virtually guarantee that soldiers will not be killed or wounded. Thus, Beauchamp and Savulescu suggest that drones are apt to make the biggest difference in the levels of support for war during humanitarian interventions that would be unpopular if they required the use of ground forces.

As these studies indicate, opinions are divided on the issue of how support for war may be affected by the type of war being considered as well as in their assessments of whether an increased willingness to fight is dangerous or morally advantageous. Testing expectation 4.1 will also allow us to evaluate the alternative relationships between drones and PPOs developed by Brunstetter and Braun and by Beauchamp and Savulescu. We can also move beyond these specific predictions to understand how the availability of drones will impact public support for other kinds of operations.

Principal Policy Objectives in American Military Operations

Early work on PPOs distinguished between wars aimed at foreign policy restraint and wars for internal political change. This distinction was made to demarcate conflicts that could have different degrees of perceived importance for national security.[9] States wage wars for foreign policy restraint when their goal is to prevent an opponent from acting aggressively, either by fighting in self-defense or in defense of an ally. Bruce Jentleson, who pioneered the study of PPOs, was careful to explain that this does not include preventive operations aimed at deterring future aggression; there must be a real and active threat that makes war a matter of national interest. By contrast, a war involving internal political change seeks to transform the domestic power structure or institutions of another state. Such wars tend to have a more pre-emptive character since there is no immediate threat to be countered but rather the presence of some more distant or indirect national security interest.

Using polling data, Jentleson shows that foreign policy restraint tends to have much higher levels of support than internal political change. This trend is particularly strong when it comes to judging a prospective military venture when the numbers of casualties and potential for success are uncertain. Jentleson reaches an optimistic conclusion based on these results: that the American public will prudently support wars against aggressive states while still being reluctant to condone aggressive or unnecessary wars. This distinction between foreign policy restraint and internal political change highlights an important difference in the degrees to which a war may be considered justified. Foreign policy restraint has a more defensive character that would more easily satisfy the demands of just war theory and international humanitarian law.[10] Internal political change is not necessarily aggressive, yet it is more difficult to justify because it goes beyond defensive war and pre-emptive war against imminent threats to take on the character of preventive war against threats that may not actually materialize.[11] Work on PPOs was cast as an intervention in an ongoing debate over what forces drive policy preferences. According to the "Almond-Lippmann consensus" that prevailed in the years following the Second World War, the American public is mercurial, lacking clear policy preferences and unable to rigorously evaluate potential courses of action.[12] The prudential considerations at work in analyses of PPOs contrasted with this view, showing that people are capable of carefully evaluating prospective actions and making consistent judgments about these.

Subsequent research by Jentleson and Britton expanded the PPO categorization to include humanitarian intervention, which they define as missions to provide emergency support for people who are suffering from a humanitarian disaster. Interventions may be aimed at not only providing protection against acts of political violence but also environmental catastrophes.[13] Jentleson and Britton argue that support for humanitarian interventions tends to fall somewhere between foreign policy restraint and internal political change. Humanitarian interventions have lower levels of support than the former because they are not seen as being as necessary for the protection of national security, but they are more popular than internal political change missions that lack the moral status of humanitarian operations.

Gelpi, Feaver, and Reifler generally substantiate Jentleson and Britton's findings, especially the conclusion that Americans make reasoned calculations about when to support war. They contend that Americans may lack detailed knowledge of what is at stake in foreign conflicts as well as knowledge of the contested areas, but that they are nevertheless capable of weighing the costs and benefits of military action when these are presented to them. As they put it, "[t]he public may not be very good at quiz bowl questions about international current events, but the public as a whole has stable and reasonable opinions that change in response to changes in the real world."[14] They challenge what they see as a misconception that members of the public cannot make reasoned decisions when they have little contextual information, and contend that Americans are good at weighing the immediate costs and benefits that are presented when politicians make the case for war.

Gelpi, Feaver, and Reifler also update the study of PPOs by pointing out the importance of including counterterrorism operations as a new form of foreign policy restraint, and they find that this type of mission receives the highest levels of support, followed by humanitarian intervention, and then internal political change. However, they argue that this ranking should not be treated as a fixed characteristic of public opinion. As they see it, the emergence of different types of threats could cause the relative interest in the PPOs to change. They speculate that policymakers' framing of threats could also influence the ordering of support for PPOs.[15] This is a critical insight. It indicates that the ranking of PPOs is not immutable but rather a reflection of the security context and more general patterns in the public's outlook toward potential threats. We may therefore expect to see shifts in PPOs popularity over time as threats and preferences change.

Some have challenged these findings, offering other explanations for variations in public support for past conflicts. One plausible alternative explanation is that attitudes are not responsive to the PPOs themselves but rather to the rhetorical framing of conflicts in terms of PPOs. By this account, some informed members of the public may be influenced by the conflict objectives while others who are less concerned with matters of foreign policy will respond to how elites justify military operations and how clearly they articulate the reasons for fighting. This means that "[w]hile well-informed citizens are likely to evaluate the policy for what it is, a majority of Americans will buy what the White House sells them."[16] This exposes the critical difference between how a conflict is framed and the real reasons why it is being waged, raising the possibility that conflicts can be misrepresented in an effort to make the case for war. Nevertheless, this finding substantiates the conclusion that PPOs are important to the public by showing that they influence support for war regardless of whether the PPOs are genuine or misrepresented.

A central finding in work on PPOs, then, is that opinions about the importance of a conflict accounts for differing levels of support and that perceptions may change over time, either as a consequence of real shifts in national priorities or in response to elite rhetoric. If this is the case, then we should expect variation in which PPOs are supported over time as the security landscape shifts and new threats emerge. Such shifts have occurred over the course of research on the relationship between PPOs and public opinion. When Jentleson first tested his theory, the United States was engaged in its Cold War mission of containing Soviet influence, a task that made foreign policy restraint appear to be an important objective even when acts of aggression where carried out against other countries. Jentleson and Britton analyzed support for humanitarian interventions in their follow-up research because of the United States' post–Cold War involvement in small wars that were meant to protect marginalized groups or to stabilize conflict zones.

In the current era, we expect that the three other PPOs will have declined in importance relative to counterterrorism operations. The threat of terrorism, first from al Qaeda, then from Islamic State, has defined US security policy throughout the period when combat drones have been employed. The link between this PPO and drone strikes is so strong that even studies making inferences about future drone operations tend to assume that these will take place in a counterterrorism context.[17] And many of the moral concerns that critics direct against drones apply not so much to the machines themselves but to the ongoing practice of using these weapons platforms to strike at suspected terrorists who, lacking advanced military hardware of their own, are powerless to retaliate.[18]

This is the logic behind our second expectation, which holds that the public will be more willing to countenance forms of military action that risk soldiers' lives—such as air strikes and the use of ground troops—when the PPO is of central importance to the national interest. We base this on the insight that casualty aversion will form a barrier against certain types of military operations and not others, and that the effects of casualty aversion will vary considerably depending on the mission. Mueller argues that casualty aversion and mission type are closely linked.[19] He contends that the American public was fairly amenable to sending soldiers into foreign countries during the Cold War because this was seen as advancing the containment strategy that appeared to be vital to the country's national security. Public support for those wars came not from the immediate objectives being pursued but from the overall goal of preventing a more serious threat from materializing, and the feeling of urgency surrounding those missions helped to maintain higher levels of support than would have been possible in costly wars that were seen as unimportant.

Smith reaches a similar conclusion. He argues that casualty aversion is affected by the extent to which a conflict advances national interests: "The explanation for the greater prominence of the casualty factor in the 1990s is simply that for democratic Western nations, most conflicts did not engage national interests at all deeply."[20] He describes humanitarian interventions like those in Somalia, Bosnia, and Rwanda as being particularly good examples of the difficulty of sustaining popular support for nonessential missions.[21] In each case, there was not a strong justification for putting American soldiers at risk of being killed or wounded, as there had been during the Cold War. Gelpi, Feaver, and Reifler show that perceptions of the importance of American military casualties in Iraq are influenced by the perceived necessity of the mission there, and in particular by the importance of Iraq as part of the War on Terror. "[W]e find that those who believe that the war in Iraq is the 'central front' in the war on terrorism are 10 percent more likely to tolerate 1,500 casualties than a respondent who believes that Iraq is a distraction from the war on terror."[22] This also indicates the importance of PPOs in general, since the disagreement between those who see a link between Iraq and the War on Terror and those who do not is a matter of whether the conflict is a counterterrorism operation or an internal political change.

Principal Policy Objective and Support for the Use of Force: Survey Experiments

Research on the reasons driving support for war typically looks at the influence of mission type on support for war using polling data collected before, during, and after actual conflicts. This provides an excellent perspective on how public opinion shifts in response to real events, yet it sometimes leaves unanswered questions when it comes to accounting for the causal mechanisms underlying those shifts. It is difficult to tell the extent to which opinion was influenced by the mission, by low casualty tolerance, or by some other factors. Experiments provide a way of more precisely determining the extent to which PPOs affect support for a war by isolating their influence and holding other considerations that might affect support for war constant. Experiments also allow us to vary the weapons and tactics used to achieve a PPO to determine its influence on public opinion relative to casualty tolerance.

To assess this chapter's two expectations, we conducted an experiment based on hypothetical news stories. The news stories varied two elements (see the chapter appendix for the complete wording of each treatment as well as the questions that comprise the survey instrument). The first was the attack type and is very similar to the experiments reported in chapter 3. The attack type could take one of three forms: a drone strike, an air strike from a piloted aircraft, or the use of ground troops. Consistent with the casualty-aversion idea discussed above, the treatments had different information about the risk that American military personnel would face. The drone treatments stated that "the use of unmanned drones means that no American military personnel would be placed at risk." The air strike treatments, in contrast, included information that the target of the strikes were believed to lack weapons capable of attacking aircraft, suggesting a low possibility of military casualties. The ground troops news stories did not mention if these troops faced any danger.[23]

The second element that varied across the treatments was the purpose or goal of the use of force and is modeled after similar treatments used in other research on public opinion and foreign policy.[24] The counterterrorism treatment described an al Qaeda branch operating in Yemen, with plans to place bombs on American airliners and in the mail. The proposed attack involving drones, air strikes, or ground forces would be directed at the al Qaeda bases with the primary goal of killing the organization's leaders before they could launch an attack against the United States. Thus, the scenario closely parallels the kinds of targeted killing operations that have become common during the War on Terror.

The treatment for foreign policy restraint described Yemen threatening to attack oil tankers passing through the Persian Gulf using missiles launched from within Yemen. This would threaten American interests, as well as the global economy, by causing a dramatic increase in oil prices.

The story is therefore a case of one state acting aggressively, with costs that could lead respondents to sense that American national security was under threat. The proposed attack would be directed at the missile installations and would not have any larger objectives, such as degrading the military or replacing the government.

The internal political change treatment stated that a pro-American government in Yemen was in danger of being overthrown by a rebel group operating within the country. It said that the rebels had already defeated government military forces in several battles and that they were nearing the capital. The American intervention was being requested by the government and would attack the rebel leaders.

The humanitarian intervention treatment described Yemen's military committing atrocities against ethnic minorities within the country. It said that the actions had been condemned by international human rights organizations and that those organizations were also calling on the United States to launch a military operation. This framed the intervention as being important on moral grounds. We were careful to not give respondents any indication that an intervention could benefit the United States or its allies or that it could lead to a protracted counterinsurgency operation. The prospective attack would only be launched against military bases housing the soldiers who were committing the atrocities, without any goals of occupying the country or reforming its government.

Combining these two elements—type and objective of military force—produces a total of twelve treatments. Roughly three hundred respondents were randomly assigned to read each of these stories. They then answered questions about their reactions to the planned use of force, including the degree to which they supported the attack, the importance of the policy objective, their estimates of the number of military casualties that would result, general attitudes regarding the wisdom of the use of force, and demographic questions such as party identification, age, gender, and so on. The experiment does not directly distinguish between support for military force (i.e., a desire to attack) and lack of opposition (i.e., an unwillingness to question the attack). However, distinguishing between levels of intensity for supporting a policy helps to address this issue by allowing those with intense feelings to strongly agree or disagree and those with weak feelings somewhat to agree or disagree as a way of signaling that they would not block the use of force.

As we pointed out in chapter 3, Yemen is a useful case for testing fictional conflict scenarios because it is in the unique position of seeming like a plausible threat even though Americans are relatively ignorant of it. It is doubtful that Yemen would threaten US shipping in the Persian Gulf, or even that it would be able to do so. However, given Americans' low level of knowledge about Yemen,[25] it is unlikely that many respondents have any sense of the country's military capacities or its interests when it comes to oil shipments. News coverage of the country identifies it as a terrorist sanctuary, but rarely comments on its military capacities or strategic goals.

It is more plausible that Iran would pose the kind of threat we describe, but identifying it as the target of our fictional strike would complicate our ability to draw conclusions about the influence of attack type and PPO on support for the use of force. Iran is a far more prominent security threat, especially given the ongoing controversy about its nuclear enrichment programs. It is also apt to raise partisan disagreements because of Democrats' efforts to rebuild relations with the country and Republicans' preference for reinstating the sanctions regime against its nuclear program.[26] This means that although people who stay informed about Middle Eastern politics may see Yemen as a less realistic setting for some of the scenarios we describe, it is a better setting from a methodological perspective because it helps to avoid the difficulties that would come from a case with a higher profile or stronger connotations.

To this point, we have argued that military casualties are an important cost of conflict that influences attitudes regarding the use of force. As with the experiments in chapter 3, we expect respondents' estimates of the number of military casualties will vary by attack type. Respondents should expect that the lowest number of military casualties will result from a drone strike, since these news stories make it explicit that military personnel will face no risk of physical harm. Treatments involving air strikes should lead to higher expectations of military casualties. Even though these conditions state that the target of the attack is not believed to have weapons capable of threatening military aircraft, respondents might still expect that the chance of military casualties could be higher. Respondents might worry that the target has, unknown to the United States, acquired anti-aircraft weapons, or that casualties could result if the aircraft were to malfunction over enemy territory. Respondents' expectations of military casualties should be highest in the treatments that describe an attack by ground troops. Although these news stories make no mention of the risks that military personnel face in such situations, it should be straightforward

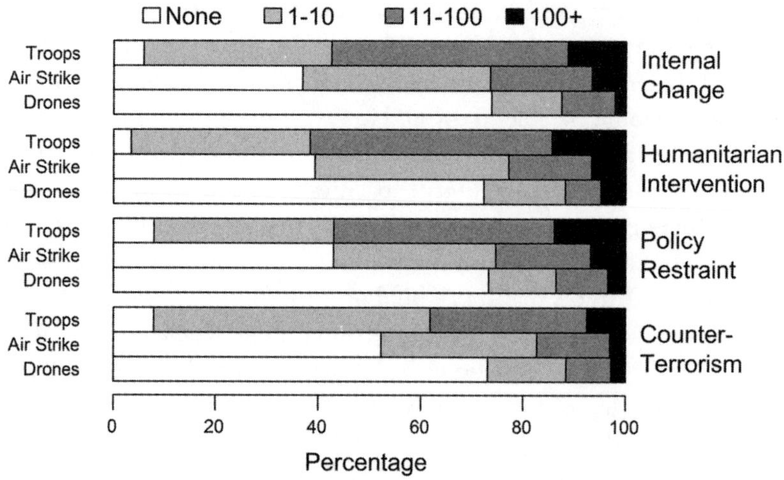

Fig. 4.1. Casualty expectations by attack type and principal policy objective

for respondents to infer such risks from the information they read.

To assess how assignment to different treatments influenced estimates of military casualties, respondents were asked if they expected no military casualties, between 1 and 10 casualties, or between 11 and 100 casualties if the attack described in the news story they read were to be carried out. Figure 4.1 summarizes answers to this question. We see a pattern, consistent with our expectations and the findings in chapter 3, in which respondents assigned to the drone strike treatments expected the fewest casualties, followed by those assigned to air strike treatments and then the ground troop treatments. Importantly, differences across treatments for the same type of attack are small. This indicates that the information in the treatments influenced respondents' expectations of military casualties in the same manner that we saw in chapter 3 and that this pattern holds across the different objectives of military action.

We next turn to assessing the influence of attack type and PPO on support for the use of force. The percentages of respondents in each treatment who selected each possible response to the question asking them to express their level of support for the attack are depicted in figure 4.2. A more systematic analysis is presented in table 4.2. Four models are reported in the table. Each uses the variable support for the attack as the dependent variable. The independent variables for each model include dichotomous variables indicating assignment to an air strike or ground troops treatment. This means that the coefficients on these variables estimate how assignment to these treatments alters support for the use of force compared to respondents assigned to the drone strike treatment. The models also include gender, education, party identification, age, and minority status covariates and are estimated using ordered logistic regression.

Two patterns emerge from figure 4.2 and table 4.2. First, consistent with expectation 4.1, we see that drones strikes garner more support than do ground troops for each PPO. This means that the effect of drones on attitudes is not limited to counterterrorism operations; instead, the technology leads to more support for *any* military objective. Second, the data is consistent with expectation 4.2, as counterterrorism operations received greater approval, regardless of the type of force being used, than any of the other principal policy objectives. Even using ground forces in counterterrorism operations, which our respondents believed created a greater risk of harm, received more support than drone strikes or air strikes in support of any of the other PPOs. Taken together, these results demonstrate that respondents were more likely to risk soldiers' lives in certain types of missions than others. In particular, there was a much higher risk tolerance when it came to counterterrorism operations than the other types of operations. This is evidence that the PPO has a strong influence on when the public will support drone strikes, or any other means of attack.

It is important to note that the pattern becomes more complicated beyond the "Strongly Approve" response, indicating that the lack of opposition shown by those with more mixed "Somewhat Approve" and "Somewhat Disapprove" responses is less consistent than the stronger feelings of support. Air strikes received lower levels of overall opposition than drones in internal political change and humanitarian intervention missions, while the levels were roughly equal for counterterrorism, and drones were less objectionable in foreign policy restraint. It is not clear why these differences exist, as they do not follow a clear pattern. It may reveal that those with weaker preferences about attack type are either less concerned about how military operations are carried out and more prone to inconsistent judgments because of their comparative disengagement. Whatever the explanation, the variation in the two intermediary levels of support does not track with the predicted casualties (i.e., there is less opposition to drones in foreign policy restraint operations even though respondents did not perceive these missions as being significantly more dangerous than the oth-

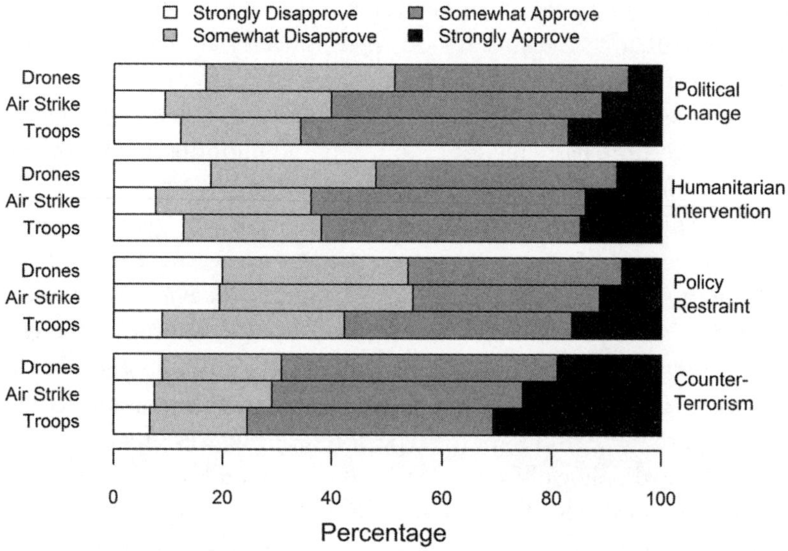

Fig. 4.2. Support by attack type and principal policy objective

TABLE 4.2. Ordered Logistic Regression Models of Attack Type by Principal Policy Objective

	Counterterrorism	Humanitarian Intervention	Foreign Policy Restraint	Internal Political Change
Air strike	–.239	–.005	–.567***	–.316***
	(.161)	(.161)	(.156)	(.159)
Ground troops	–.463***	–.689***	–.768***	–.777***
	(.160)	(.169)	(.153)	(.160)
Gender	–.409***	–.171	–.473***	-0.173
	(.134)	(.134)	(.127)	(.131)
Education	–.142	–.022	–.414	–.812
	(.314)	(.314)	(.278)	(.306)
Party identification	1.183***	.701***	1.392	1.12***
	(.218)	(.224)	(.211)	(.226)
Age	.02***	–.011***	–.004	–.001
	(.006)	(.006)	(.005)	(.006)
Minority	.223	–.119	.082	.062
	(.156)	(.165)	(.149)	(.152)
Likelihood ratio	1,908	1,908	2,130	1,999
AIC	1,928	1,928	2,150	2,019
Observations	833	806	870	854

Note: Standard errors in parentheses; * $p < .1$; ** $p < .05$; *** $p < .01$.

ers). From this we can at least conclude that the lack of opposition is probably not being influenced by casualty expectations.

To this point, we have found that attacks from drones receive more support than do attacks involving ground troops across different PPOs and that respondents are more willing to risk incurring casualties for counterterrorism objectives. But how large is this effect? Does the availability of drones sharply increase support for the use of force? Using the regression models from table 4.2, we calculated the predicted probability that a respondent would select each of the four levels of the dependent variable measuring her or his support for the attack carried out by a drone, holding all other variables constant for each PPO. We then repeated this exercise for respondents whose treatment assignments described the attack as being carried out by ground troops. Subtracting the latter value from the former provides an estimate of the difference in the predicted probability of each level of the dependent variable for respondents in the drone and ground troop treatments. Figure 4.3 graphs these changes for each PPO. For each PPO, respondents were more favorable to attacks by drones than by ground troops. However, there are noticeable differences in the size of these effects across PPOs. For the internal political change, foreign policy restraint, and humanitarian intervention treatments, reading about an attack by ground troops reduces the likelihood that a respondent will "strongly agree" or "somewhat agree" with the decision to launch the attack by between 15 and 20 percent. The effect of an attack by ground troops compared to drone strikes was about half this size for respondents who read about a counterterrorism operation. In other words, while respondents in the counterterrorism group also preferred drone strikes to the use of ground troops, they were considerably more willing to endorse attacks with troops.

The chapter's second expectation offers two reasons for the comparatively high levels of support for counterterrorism operations. First, terrorism has been the most prominent threat to national security since 9/11, with new terrorist plots being continually uncovered and governments around the world making a concerted effort to kill or capture suspected terrorists before they can strike. Even conventional military operations in Afghanistan and Iraq have been framed as contributing to the Global War on Terror.[27] The importance attached to counterterrorism operations probably had a profound influence on respondents' conceptions of national security priorities.

Second, the counterterrorism scenario was the only one that involved

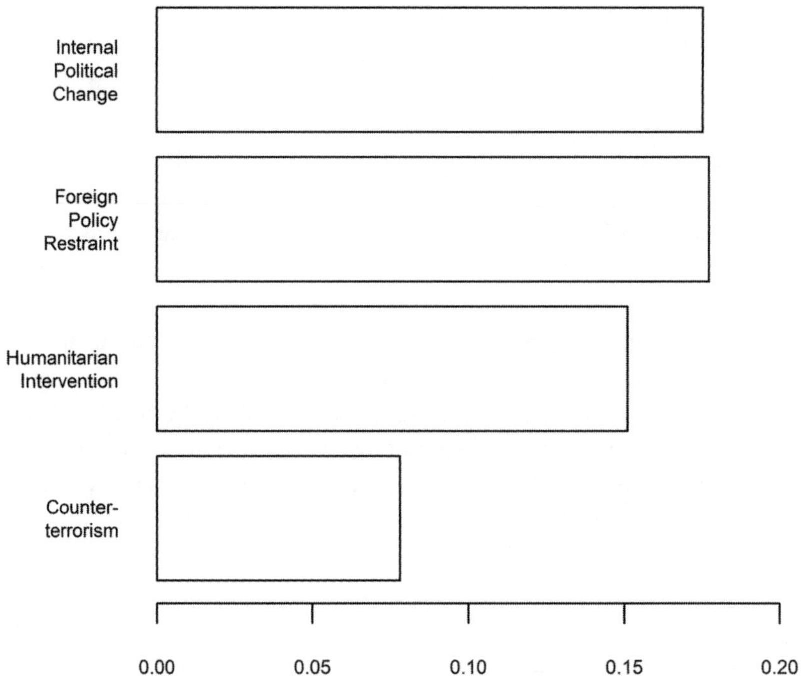

Fig. 4.3. Ground troop changes in predicted probabilities of supporting attack

an immediate threat to American national security. The fact that planes and the mail service would be used as the delivery vehicles for an attack would have been particularly threatening given terrorists' previous use of these methods to carry out attacks within the United States. To determine if this is the case, we asked respondents the following question: "The United States faces many challenges today, at home as well as overseas. Among these challenges, how important do you think it is to address the problems discussed in the news story you read earlier?" Responses for each treatment are depicted in figure 4.4, where we see that considerably more respondents identified all of the counterterrorism objectives as very or somewhat important than they did for the remaining PPOs. Looking more closely at the patterns of coefficients in table 4.2, we see that respondents assigned to the counterterrorism and humanitarian intervention treatments offer as much support for the planned operation when it employs air strikes as when it uses drone strikes. This indicates that these respondents were

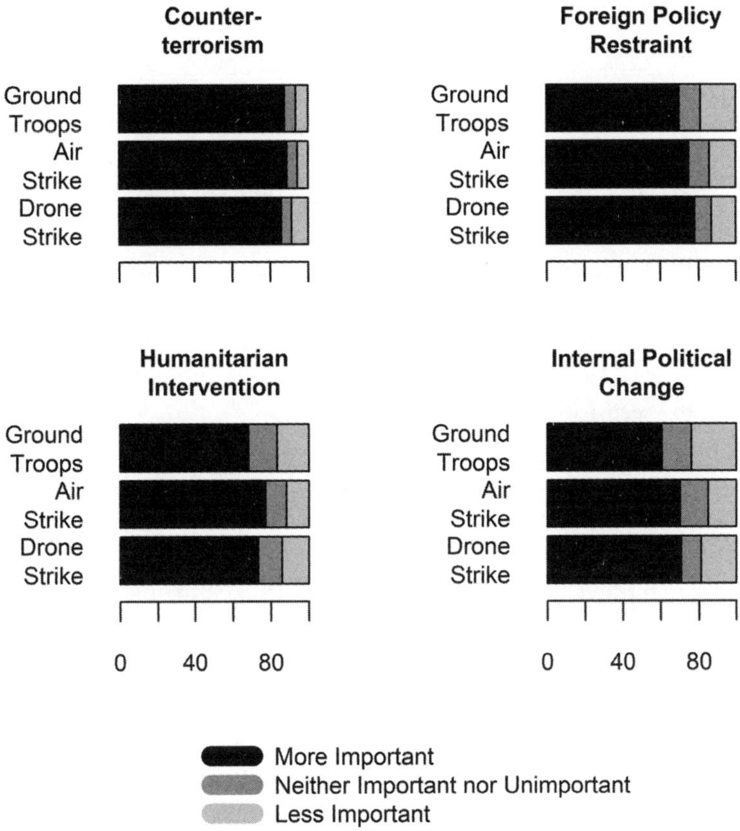

Fig. 4.4. Support for attack

willing to place air crews in some danger to achieve these objectives, but on average were less willing to do so when the objective involved internal political change or foreign policy restraint.

There was more overlap in the levels of support for the other types of operations. We did not see the large differences in support across the foreign policy restraint, humanitarian intervention, and internal political change treatments that one might expect from Jentleson and Britton's work on PPOs. This may be due to the way we described the objectives. The internal political change mission was described in ways that made American intervention appear to be benevolent—an effort to help a beleaguered ally against rebels. Had the scenario described the rebels in favorable terms or provided some indication that the government being protected was corrupt or abusive, then we would probably have seen a significant drop in the levels of support.

Support for internal political change might also have been lower if the mission was to depose an existing government, as this might carry more risk of escalation and be seen as less politically defensible than efforts to protect a government against rebels. The possibility for variation here again highlights the importance of being attentive not only to the differences between PPOs but also to the various reasons why each PPO might be pursued. Even when focusing on a specific type of PPO, the perceived moral and legal legitimacy, as well as the perceived centrality of the mission to national security, could vary considerably.

Even though our results differ slightly from Jentleson and Britton's findings, our experiments confirm those authors' claim that the American public does calculate the costs and benefits of prospective policies. The degrees of support coincided with the severity of the threat to national security identified by the respondents in our experiment. The counterterrorism scenario introduced a clear and plausible threat, while the next most popular intervention was directed at the less immediate, though still very serious, threat of economic disruption. The internal political change scenario did not describe any benefits to American security, yet the act of helping a foreign government did suggest that the United States might be looking out for its interests by building or strengthening relations with an ally. Finally, the humanitarian intervention not only failed to describe any benefits but also put the United States on the side of marginalized people who would have little to offer as allies.

The low approval for humanitarian interventions regardless of the weapons being used is evidence that casualty aversion is not the only obstacle preventing states from getting involved in these types of conflicts. The risk of sustaining casualties certainly plays some role in shaping support for humanitarian interventions, as evidenced by the lower levels of support for committing ground forces than drones or manned aircraft, yet it is clear that even relatively low-cost interventions are unpopular. This supports Jentleson and Britton's contention that conflicts that are less essential to national security will generally receive less support and that the public will not favor military operations that seem unnecessary.

The pragmatic attitude toward wars is important for evaluating the consequences of casualty aversion discussed in chapter 3. It is revealing that research on casualty aversion and criticisms of warfare tend to reach much different conclusions from their shared belief that citizens of contemporary democracies have a low tolerance for costly wars. Most critics of drones see casualty aversion as a positive development that discourages war and provides greater democratic accountability. By contrast, other commentators regret this phenomenon because it limits the United States' capacities for deterring enemies and for launching attacks when necessary.[28]

The difference between these attitudes seems to be rooted in different conceptions of how and when the United States will wage wars. Those who are concerned about casualty aversion interfering with American foreign policy or who think it is not a serious barrier against fighting often seem to envision American wars as actions taken against unjust aggressors or as the military rescuing beleaguered populations during humanitarian interventions. By contrast, critics of drone warfare strongly imply that wars should be avoided and focus much of their attention on abuses of US military power. They are particularly sensitive to immoral, illegal, or politically inexpedient actions performed during the War on Terror.[29]

Future conflicts fought using drones will doubtless take many different moral valences. Some may be unjust, aggressive wars. Others may be benevolent or defensive. Most will probably be morally ambiguous conflicts the character of which will be continually debated, which seems to be the norm for most conflicts. There is rarely a consensus on when America's wars are justified, even among just war theorists.[30] The inevitability of different types of conflicts with divergent moral and legal implications suggests that pilot invulnerability will not be simply good or bad but that it may instead assist in achieving PPOs that differ in their moral character and political importance. Greater support for drones compared to other attack types may be desirable or undesirable depending on the circumstances of a particular conflict. The morality of drone warfare will rest heavily on how drones are used and when, which puts a great deal of responsibility on the elites who choose when to launch attacks. It also places responsibility on the public, which will have to make decisions about when to support the use of military force.

In practice, the shifts in public opinion of this magnitude are alone likely to decide whether a particular war is waged only under certain treatments. Decisions about initiating wars are not made directly by the American public but rather by elected officials whose decisions may be insulated from or unresponsive to citizens' attitudes. Thus, public opinion will matter to the extent that changes in it can influence policymakers and alter their decisions. Voters may decide to punish policymakers who wage unpopular wars by removing them from office.[31] However, it seems unlikely that any shifts in public opinion that could be caused by using drones instead of ground forces or piloted aircraft will have great influence on policymakers' decisions if they are determined to go to war. A decision that increases opposition is probably unlikely to affect a policymaker's election prospects, especially when it is weighed against other decisions that person has made.[32]

The effect that drones have on support for war is likely to make them attractive weapons for politicians who are concerned with sustaining their popularity or who are not strongly committed to the use of force. The shift in public support that drones produce may not be enormous and may not be sufficient to change public opinion about a war, yet it is nevertheless just one of the many advantages that make drones attractive weapons. As Sauer and Schörnig correctly point out, there are multiple reasons for preferring drones over other weapons and tactics, such as their ability to loiter over targets and their comparatively low cost.[33] Nevertheless, based on our experiment, it appears that the predictions about drones undermining democratic accountability are probably too strong. Because drones produce moderate increases in support for war, greater reliance on them will probably be unable to silence anti-war voices. This is evident from figure 4.2, which shows that roughly 20 to 40 percent of respondents did not want to launch drone operations despite the invulnerability of soldiers operating them. Such consistent opposition to military operations from that many respondents is evidence that politicians will continue to have strong incentives to pursue peaceful strategies of conflict resolution in an effort to satisfy those citizens.

Conclusion

This chapter takes a step toward building a more sophisticated understanding of how drone technology shapes support for war by exploring the influence of principal policy objectives. Accounting for the divergent levels of support across PPOs is useful when assessing what effects drones will have on support for future military operations and whether these weapons platforms will allow policymakers to escape some of the constraints arising from casualty aversion. Furthermore, the differences between the support for different PPOs in our study compared to others shows that there is scope for variation within each PPO depending on what a specific operation aims to achieve. This all points to respondents making reasonably prudential judgments about prospective military ventures, granting or withdrawing support on a case-by-case basis.

Our results in this chapter indicate that attacks with drones receive more support than does the use of ground troops and, in some PPOs, air strikes. Choosing drones over these alternatives can offer enough of a boost in support to make drones popular for policymakers and to possibly even tip the balance of public opinion when there is a narrow divide between pro- and anti-war positions. But we also caution against overstating the effect of this finding. In many instances the added support that can be gained by choosing drones will not decide whether the country goes to war. Drones may therefore help to evade the effects of casualty aversion, but may not alone produce any serious changes to civic engagement or grant policymakers free rein over military affairs. Moreover, our results show that a substantial number of respondents oppose fighting regardless of whether the risks to soldiers can be eliminated. In the following chapter we will build another level onto this account of what drives support for drone strikes by looking more carefully at whether variations in the likelihood of mission success matter and whether there is a "moral hazard" of drones making it too easy to fight wars when the chance of victory is low.

Throughout this discussion we have also pointed to the issue of framing. Our experiments give respondents the information needed to assign prospective operations into a particular PPO, yet there is no guarantee that framing by media elites or policymakers will be accurate. There will always be a risk of a less popular mission type being misrepresented, such as foreign policy restraint being characterized as essential for counterterrorism. Given the effect of PPO on preferences for war, it is vital to be attentive to how conflicts are framed and to critically evaluate the case for fighting.

How elites frame a conflict, then, could influence the effect of drones on support for war. "Elite cues" approaches to public opinion hold that people generally respond to wars in ways that mirror their party affiliations: "partisan political actors, not ordinary citizens, balance costs and benefits when deciding to lend support to military action. The public may appear 'rational,' but only by following elites who share their basic political predilections."[34] This would mean that people generally do not form their own opinions about when wars are justified. Rather, they simply adopt the views of elites, who are really the ones that make decisions about when to fight.

We have not sought to assess the role of elite cues here, but it is possible to draw several conclusions about the effect cuing has on support for drone strikes in particular. First, if Americans are responsive to elite cues and have fairly low competence for making their own decisions about when to support or oppose wars, then this is a problem that predates the widespread use of drones. Berinsky finds support for this theory in evidence gathered during the Second World War, the Korean War, the Vietnam War, and the First Gulf War, as well as the early years of fighting in Afghanistan and Iraq. These cases all predate drone warfare. Second, if public support for war is largely unaffected by casualty sensitivity or PPOs, then the general conclusions we have reached would still hold true. Drone warfare should not have a major influence on the incidence of war; casualty aversion would not be able to restrict policymakers' war-making powers nor would it be reduced by the use of drones; and support for using drones would be highest when policymakers cue their supporters to take a particular threat seriously, as they have done by emphasizing the threat of terrorism.

The pattern that is evident in the levels of support for using drones, piloted aircraft, and ground forces shows that there is a continuum of support for military force that extends across the range of weapons and tactics that may be employed. One possibility for future research is to include a more diverse assessment of the weapons and tactics used by the US military to see how much this pattern is sustained. There may be gradations of support between the three types of attacks we explore. For example, support for the use of special operations forces could fall somewhere between support for an attack involving conventional ground forces and an air strike. Alternatively, it is possible that support for war could further diminish or increase as other weapons and tactics are introduced. The deployment of large numbers of soldiers to directly engage in combat could be less popular than the deployment of smaller numbers of soldiers serving in advisory roles. One could also argue that the PPO theory and casualty-aversion theories are incorrect, or at least that they are not the primary forces shaping attitudes toward wars.

Because drones have only been used on a large scale in counterterrorism operations, it can be difficult to separate their use in the War on Terror from the potential uses they might have in other types of conflicts. This is reflected in the extensive literature on drone use discussed in chapter 1. Critics often condemn the use of drones to conduct targeted killings as though these were inherently linked; the circular logic uses the evidence of American military overreach in the War on Terror as evidence against drones and uses the proliferation of drones as evidence that the War on Terror is out of control and counterproductive.[35] The concern that drones may lower the threshold for war is also closely associated with the War on Terror, as drones have facilitated attacks against suspected terrorists in Yemen, Pakistan, and Somalia, attacks that might not have taken place if drones were not available.

It is impossible to predict how drones will be used in future conflicts, especially those involving PPOs other than counterterrorism. However, based on our experiments, it seems that the American public will probably support the use of drones in other types of operations, including attacks against state military forces. It is also likely that unless there is a marked shift in the kind of PPO that is seen as most necessary for the protection of national security, the government will not have quite as much freedom of action when deploying drones for reasons other than counterterrorism. While counterterrorism operations involving drones receive relatively high levels of support, the use of drones to attack foreign states or to strike rebel militias could create stronger opposition.

CHAPTER 5

Drones and Moral Hazard

The United States launched hundreds of drone strikes against suspected terrorists in at least seven countries between 2001 and 2016. Government officials have justified these attacks on the grounds that they were essential for preventing terrorism, and they argued that no peaceful alternatives were available. Their goal was to demonstrate that clear and imminent threats compelled defensive military action and that drones' selectivity made them a less costly means of attack than the alternatives. However, critics challenge these claims, arguing that the attacks failed to achieve their strategic objectives, that less violent options were available, and that confidence in these new weapons platforms has led the United States to fight unnecessary wars that may ultimately produce more terrorists than they eliminate. These arguments highlight the possibility that drones have allowed the United States to wage counterproductive or unnecessary wars that go beyond what is necessary for self-defense—potentially intensifying the War on Terror and destabilizing international security. In particular, they indicate that the United States may be using drones to launch military strikes even in instances when nonviolent diplomatic alternatives still have some hope of success or could even be more effective.

Over the previous chapters we have considered some of the ways drones could influence support for military operations, and we have shown that many different factors may jointly shape public opinion. We have also raised some of the most plausible arguments for thinking that drones could lower inhibitions against fighting and argued that an understanding of the roles of casualty sensitivity, principal policy objectives, and the means of attack is essential to evaluate when Americans support drone warfare. These issues usually take a central place in explanations of when Americans support wars. Nevertheless, we have thus far omitted several key details that are essential for situating prospective attacks in the larger political and strategic context: whether the attacks will succeed and how the availability of nonviolent alternatives may alter opinions. In this chapter we add another layer to our account of the forces shaping attitudes toward drone warfare by taking up the issue of how expectations of mission success influence Americans' willingness to launch attacks.

Previous research has shown that anticipations of success may have a powerful influence on determining overall support for military actions. Gelpi, Feaver, and Reifler, in particular, have sought to demonstrate that anticipated success shapes the American public's willingness to support conflict.[1] Although they do not test this claim with reference to drones specifically, their research indicates that anticipated success could increase enthusiasm for drone strikes, just as it seems to increase enthusiasm for military operations in general. Drones' selectivity could raise the predicted chances of success if it leads individuals to conclude that drones are more effective than the other weapons or tactics that could be used. This would be a reasonable inference given the efforts President Obama and other senior American officials made to present drones as ideal counterterrorism tools. Thus, the use of drones could give the public greater confidence in victory if the selectivity these machines offer seems to increase the odds of winning while also decreasing the potential costs of fighting. Increasing efficacy while reducing risks may be the kind of win-win scenario that would make drone strikes far more attractive than alternative means of fighting—and possibly even more attractive than nonviolent methods of conflict resolution.

This relationship between drones and predicted success—the possibility that the perceived effectiveness of drones could increase the public's willingness to use them and that increased odds of winning could build support for more wars—leads to one of the most serious potential risks associated with drone warfare. Kaag and Kreps advance the idea that drone technology creates what they term a "moral hazard." A moral hazard exists when a person may be tempted to take greater risks because the associated costs of failure are low. Because drones permit force to be used with no chance of military casualties, their availability should also increase an individual's support for "risky or morally questionable behaviors because he or she does not have to face the consequences of these actions."[2] Drones may therefore allow states to engage in wars that are less likely to succeed simply because the harmful consequences of fighting can be shifted onto foreign populations.

This same concern is echoed throughout critiques of American drone strikes, often with the help of other concepts that also point out that drones' increased selectivity could paradoxically cause an escalation of violence. As we show, the threat of moral hazard is probably the most pervasive moral critique of drone warfare. The widespread appeal of this argument makes it an important subject for experimental research. In this chapter we contribute to research on the relationship between mission success and support for war, and we evaluate the plausibility of drones creating a moral hazard with an experiment focusing on how support for attacks changes depending on what alternative approaches to conflict resolution are available. Testing support for war in this way makes it possible to gauge whether people become more aggressive, or at least more willing to incur the costs of fighting, when the adverse consequences of war can be minimized by drone technology.

Table 5.1 summarizes our expectations regarding how drones influence estimates of battlefield success and moral hazard. Our results show that predictions of success are relatively stable across different methods of attack. Regardless of whether drones really are more or less effective than other types of military force, the American public does not perceive them as being any more or less effective. Even more importantly, we find that drones do not appear to create a moral hazard. Respondents showed roughly the same levels of support for military operations involving drones, piloted aircraft, or ground forces as the availability of nonviolent approaches for defusing a conflict changed. Respondents did not, as the moral hazard argument would suggest, become more willing to launch unnecessary strikes simply because drones were available to conduct these at a low cost to American military personnel. This undercuts one of the strongest and most commonly cited reasons for opposing drone warfare. The threat of moral hazard is therefore overstated in arguments against drones, at least when it is framed in terms of the American public's risk preferences.

TABLE 5.1. Mission Success Expectations

Expectation 5.1: Drones will lead individuals to believe they are more likely to achieve their battlefield objectives than other types of military force.
Expectation 5.2: Support for an attack that takes place despite the availability of peaceful alternatives will be higher when drones are involved.

Success and Support for War

Gelpi, Feaver, and Reifler's research coincides with our findings in previous chapters that casualty aversion and PPOs matter when it comes to shaping public opinion, but they contend that expectations of success play an even larger role in determining attitudes toward military operations. They argue that the American public is willing to support costly wars if there are grounds for thinking that the war will achieve its goals. Experiments they conducted on support for hypothetical interventions in Yemen revealed that "moderately high prospects for success may be sufficient to maintain public support, while casualties did not curtail support until they reached fairly high levels."[3] This demonstrates that casualty sensitivity does influence public opinion, but that this and other causal factors can actually be outweighed by the lure of victory or the fear of defeat.

Although they do not test public support for drone strikes specifically, Gelpi, Feaver, and Reifler's work indicates that expectations of whether an attack will achieve its objectives are important to bear in mind when testing attitudes toward drones. The desire for success appears to be a general preference that remains consistent regardless of what specific weapons or tactics are employed. But if certain weapons or tactics are more effective than others (at least in the sense that they raise the perceived likelihood of success), they could help to bolster public support. Their research suggests that the ability drones have to reduce American military casualties may be superfluous, or at least unnecessary, when it comes to building public support to wage a war if policymakers can demonstrate that they make success more likely. That is to say, leaders may continue to employ ground forces in future attacks and forgo the use of drones provided there is confidence in victory and losses are kept to a tolerable level. Conversely, their results also raise the possibility of drones being used to maintain support for military operations when a good outcome seems unlikely. This makes it crucial to be able to distinguish between the effects of casualty aversion and evaluations of mission success and to compare the relative weight of these causal mechanisms when formulating a theory of support for drone strikes.

Anticipations of success simplify the temporal issues we confronted in previous chapters. Recall that there is disagreement on the issue of whether casualty aversion emerges primarily during a war and builds as the costs of fighting are realized, or whether it is an anticipatory phenomenon that prevents or limits prospective military operations. The former deals with real costs of fighting while the latter depends on predictions. When it comes to the anticipations of success, judgments are always predictive. Regardless of whether they come before or during a conflict, they require an imaginative leap into the future. There may be more evidence based on information from past wars, yet even then the judgment retains its future orientation. As the authors say of the Iraq War, "respondents' beliefs about whether the U.S. would succeed in Iraq tend to be prospective in nature. The retrospective judgments about weapons of mass destruction and Hussein's links to al-Qaeda were not causes of respondents' beliefs about the likelihood of success."[4] By positing anticipated success as the decisive influence on support for war, Gelpi, Feaver, and Reifler provide a compelling way of explaining public opinion regardless of how far a conflict has progressed or whether it has even started. This simplifies the search for causation and makes it possible to gauge support for prospective and ongoing wars using the same experiments.

One challenge that Gelpi, Feaver, and Reifler's research raises is determining the extent to which predictions of sustaining casualties and predictions of success overlap. It is not immediately clear whether people think minimizing casualties is part of what constitutes success, or whether expected casualty numbers are distinct from the criteria that are used to evaluate success. They are aware of this challenge and in particular the counterargument they face from those who might say that the American public is apt to gauge success primarily, or even entirely, in terms of minimizing American military casualties. However, they maintain that this did not affect their results. As they explain, "the proportion of respondents who stated that the number of U.S. casualties was their measure of American progress toward success was less than 5 percent across eight separate waves of surveys."[5] Based on this, we can surmise that expectations of success are not identical to expectations of incurring casualties and that these two issues should be distinguished as we search for the causal processes that drive support for military operations.

Other work has also shown that perceptions of success shape public opinion. Gartner tests the relationship between predictions of success and casualty estimates and finds support for Gelpi, Feaver, and Reifler's position. He discovers that anticipations of the outcome influence Americans' willingness to fight and that this is independent of casualties. He confirms this relationship by only mentioning casualties late in a survey experiment, after gauging the predicted outcomes.[6] This offers additional support for thinking that anticipated success is a distinct causal variable that is largely independent of casualty sensitivity and that may therefore produce different effects.

Berinsky concludes that success does matter when it comes to determining support for fighting, but he says that the perception of success "is heavily influenced by partisanship."[7] He challenges Gelpi, Feaver, and Reifler's contention that success matters and argues instead that attitudes toward war are largely dependent on how elites work to mobilize public opinion. It is therefore the elite cues that make success or failure salient.[8] Agreement among elites raises the public's willingness to fight, while partisan divides cause sharp differences in public opinion. In another study, Berinsky and Druckman contend that this raises the possibility of reversing the causal direction that Gelpi, Feaver, and Reifler identify, making support for war the cause of anticipated success, rather than anticipated success the cause of support.[9]

Gelpi has since defended the thesis he and his coauthors advance with experiments showing that respondents exposed to news stories about Iraq were highly susceptible to the good or bad framing of events that indicated the potential for an American victory.[10] He did note that partisan identity helped to inform starting levels of support and the extent of the change in opinion, though with the prospect of success still playing the central explanatory role. There is good reason to think that partisan identification and perceptions of success both matter when it comes to shaping the overall levels of support for war, but because the former appears to influence the latter, anticipated success remains the proximate cause that is more directly guiding public opinion. The weight of the evidence therefore demonstrates that predicted success is apt to shape public support for military operations in general and possibly for drone strikes specifically, regardless of what prior conditions influence predicted success.

How Drones Influence the Prospects of Success

The importance of predicted success in the public opinion literature provides a clear link between this literature and research that is focused on drones, as one of the central points of disagreement in research on drones is whether they are effective. Proponents of drones contend that these weapons platforms are more effective than other weapons and vehicles because they can launch highly selective attacks, even against individuals and moving targets that would be difficult for long-range missiles or piloted aircraft to target.[11] They employ relatively light weaponry compared to manned aircraft and are able to strike much more quickly than long-range precision weapons, eliminating delays between a missile's launch and its impact. Quick and light strikes help to minimize the extent of destruction when drones are used to attack, but their selectivity also depends heavily on their proficiency with the kinds of preattack surveillance operations for which they were originally designed. Drones are able to track and monitor potential targets for hours or days to ensure that their weapons are directed against enemy fighters. They are even used to track the "patterns of life" of known terrorists, which can then be used to locate new suspects who follow similar patterns.

Selectivity is particularly important when it comes to avoiding harm to civilians and their property. Many of the arguments in favor of drones rest on the claim that removing soldiers from the battlefield could make it possible to strengthen norms and laws governing the use of force. The reasoning is that armed forces that do not risk their personnel in combat gain greater control over how and when they fight, which amounts to improved opportunities for selective attacks that are only carried out when the risk to civilian bystanders is low.[12] For example, Byman says that "drones, unlike traditional airplanes, can loiter above a target for hours, waiting for the ideal moment to strike and thus reducing the odds that civilians will be caught in the kill zone."[13] The selectivity and pilot invulnerability of drones may therefore come together as complementary advantages that enable unprecedented compliance with civilian immunity to coexist with improved lethality.

Aside from the obvious moral and legal reasons for preventing attacks on civilians, increased precision may also generate more pressure on government officials and military personnel to avoid civilian casualties. Beier argues that improvements in weapon precision have generated higher expectations that civilians and civilian infrastructure will be spared during attacks and that the public will not tolerate indiscriminate violence, a point to which we return in the next chapter.[14] Shane reaches a similar conclusion and contends that drones may be partly responsible for sensitizing the public to civilian casualties with promises of being able to target individual enemy combatants.[15]

These assessments appear to be borne out by the debate over drone strikes. Critics blame drones for inflicting as many as 800 civilian casualties in Pakistan, Yemen, and Somalia, though some cite numbers much higher than this.[16] Even the high estimates tend to be relatively small compared to those from previous conflicts or recent ground combat operations in Afghanistan and Iraq. According to the US military's estimates, around 66,000 civilians were killed in Iraq between 2004 and 2009.[17] This is far more than the losses inflicted by drones throughout the Bush and Obama presidencies. Moreover, the military's Iraq casualty estimate is lower than many of those offered by independent sources,[18] suggesting that the true gap between casualties inflicted by ground forces and drones could be much larger. Vocal opposition has plagued drones since they were first armed even though they inflict low casualties compared to ground operations like the one in Iraq, which lends support to the idea that increasing precision and reducing civilian casualties is not enough to mollify the public.

The selectivity of drone strikes may be especially advantageous when it comes to creating the conditions for lasting peace in destabilized areas. One of the most serious challenges in counterterrorism and counterinsurgency is defeating committed enemy fighters while also gaining the support of civilian populations.[19] The American counterinsurgency doctrine unveiled in 2006 prioritizes the goal of cultivating indigenous support and enlisting local people as allies in the war effort, or at least ensuring that they withhold support from opponents.[20] Drones may plausibly advance this goal if they are able to strike at enemy fighters without harming civilians or damaging their property. After all, limited strikes that only target enemy combatants are apt to have a more modest cost to America's image in contested areas than more aggressive operations that inflict heavy casualties.

Drones not only hold important advantages when operating independently but could also improve the effectiveness of other units when used in conjunction with them. The relatively low costs of building and maintaining drones, as well as the absence of onboard pilots, make it possible to carry out attacks against well-defended targets that may be too dangerous for manned aircraft to pursue. Existing drones may be too vulnerable to air defense systems to be useful in this role,[21] but as future generations of these aircraft become more advanced they might be used alongside manned aircraft and sent on hazardous missions without placing pilots at risk. Drones' ability to stay airborne for long periods of time would make them ideally suited for supporting ground forces with accurate intelligence and close air support. They could also help to coordinate other support units, such as by locating targets that can be attacked by artillery or manned aircraft carrying heavier ordnance. For now, this is only speculative, but it is a plausible future scenario considering the popularity of tactical reconnaissance drones among American ground forces and the money that has been invested in exploring ways that combined-arms operations could be conducted.[22]

Advocates characterize drones as an invaluable means of eliminating key members of terrorist organizations, especially those in leadership roles, and rendering the organizations incapable of functioning.[23] This strategy may be a way of eliminating terrorist threats without sending ground forces to occupy foreign territories and possibly without even killing lower level members of terrorist organizations. In 2013 President Obama defended the strikes by saying that "[d]ozens of highly skilled al Qaeda commanders, trainers, bomb makers and operatives have been taken off the battlefield."[24] He went on to emphasize that these people were all imminent threats who had to be killed for national self-defense. "Plots have been disrupted that would have targeted international aviation, U.S. transit systems, European cities and our troops in Afghanistan. Simply put, these strikes have saved lives." The administration's defense of drones has consistently depended on this argument that they succeed in defending the United States and its allies against terrorists who would have otherwise carried out deadly attacks.

Opponents of drone warfare have countered these arguments by attempting to show that drones are generally ineffective and may even exacerbate military and political challenges. Some counterinsurgency theorists have argued that despite their capacity for carrying out precise attacks, drones are problematic because they alienate potential supporters. David Kilcullen and Andrew Exum contend that drone strikes are counterproductive when used to carry out targeted killings against terrorist leaders.[25] They argue that drones create a "siege mentality" among foreign civilian populations, generate outrage in the areas where they are used, and allow technology to replace strategy. The first and second processes result in declining support for the United States and reduced willingness to cooperate in the future. The third process leads to a decline in military effectiveness that could make it more difficult for American forces to win the War on Terror. These conclusions about the ineffectiveness of drones have been taken up by many critics of drones and targeted killings more broadly.[26]

They have gained a special weight because of the authors' contributions to forming United States military policy. Kilcullen is one of the leading theorists behind the style of population-centric counterinsurgency operations the United States military conducts and was an advisor to General Petraeus, while Exum is an army veteran of the wars in Iraq and Afghanistan.

Michael Boyle raises even more serious strategic and political objections to drone warfare, arguing that the use of drones in countries that are struggling against domestic insurgencies, such as Yemen, Pakistan, and Somalia, undermines those states' legitimacy and prevents them from developing more effective security institutions. He reasons that "drones provide a powerful signal to the population of a targeted state that the perpetrator considers the sovereignty of their government to be negligible,"[27] which in turn builds domestic opposition to the state and generates greater support for terrorist or insurgent organizations. By this account, drones may be counterproductive under many conditions, especially in the type of counterterrorism scenario that received the most support in our experiments in earlier chapters. Other commentators have argued that relying on drones will increase military asymmetries, thereby forcing any belligerents on the receiving end of drone strikes to resort to terrorism in a desperate attempt to survive.[28]

It is unclear whether these assessments of the strategic effects of drone use are correct, and indeed this is beyond the scope of our research on how drone warfare interacts with public support for war. However, the debate over whether drones help or hinder the pursuit of military and political objectives raises the question of what the American public *expects* to happen when drones are used and how the expected effects could feed into use of force preferences. Predictions of success in drone strikes compared to other types of attacks will provide deeper insight into the public's overall attitude toward this type of warfare and whether drones strikes are perceived as affecting the outcome of conflicts. This leads us to our first expectation for this chapter: the American public will believe that drones are more likely to achieve their battlefield objectives than are other types of military force.

The Effects of Moral Hazard

Expectations of success or failure in military operations are not mere guesses at what the future may bring. They affect present decisions, possibly in ways that change the criteria for when military force is warranted. It is here that the effects of moral hazard need to be considered. The concept of a moral hazard has its origins in the insurance industry, and it is worth briefly considering how it developed to appreciate its mechanisms and its potential consequences. Originally the term referred to the possibility that having coverage for a particular type of harm might make people less vigilant in guarding against it. As Rubinsten and Yaari explain, "the insurer cannot observe certain actions taken by the insured" and this "creates an incentive for the insured, once insurance is purchased, to act in a matter that is liable to enhance the likelihood of a large claim being filed."[29] As they characterize it, the problem follows from the challenge of observing those who buy insurance. Any status or behavior that was recorded before purchasing insurance may change as a result of being insured, invalidating the initial estimates. This casts the problem from the insurer's perspective. For the insured, the problem is one of feeling freer because of the security of shifting the risk onto someone else. Here the problem is not so much about monitoring as it is about displacing responsibility. Seen from either perspective, the moral hazard is potentially dangerous because it leads the insurer into a riskier investment and the insured into riskier behavior.

The concept of risk plays an important role in creating a moral hazard, as it essentially refers to transference of risk from one party to another and the possibility that risks may become magnified because of the insured party's relative freedom from negative consequences. Information asymmetries are also important to creating moral hazards. The person who takes insurance coverage may be unaware of the costs that carelessness may have on the insurer, while the insurer may not be able to predict how the risk calculus will change as a result of the insurance policy. Evaluating moral hazard therefore demands a careful look at the forces shaping risk evaluations and the mechanisms for oversight that could potentially prevent a more risk-acceptant stance from emerging as the costs of one actor's mistakes are displaced onto another.

Interest in moral hazard has spread far beyond insurance to become a more general challenge of managing the behavior of individuals and organizations. The concept has become especially prominent in economics, giving rise to countless applications. For example, there is the moral hazard that financing a company will give it greater freedom to make irresponsible choices or that deeming a bank is "too big to fail" will make it more confident in issuing risky loans.[30] There are ways of limiting the effects of moral hazard experienced by insurance companies, such as with the improved monitoring provided by car navigation computers or more rigorous follow-up procedures.[31] The insurance business has progressed by finding ways to reward the insured for tracking behavioral changes, but that strategy may be difficult to replicate in other domains. It is particularly challenging to monitor this phenomenon as it is manifest in the collective decisions made by entire organizations, decisions that can be inherently difficult to identify because of the multiple actors involved.

A similar process affects technological innovations through what is known as the Peltzman effect. This name comes from Sam Peltzman, who documented the phenomenon in car safety regulations.[32] Improvements in a machine's safety may cause corresponding changes in how a machine is used, and this may ultimately make the machine more dangerous. For example, improvements to car seatbelts, brakes, or steering can lead drivers to feel more confident, thereby increasing their willingness to drive dangerously. In one book on the subject, Greg Ip blames everything from financial crises to car crashes on the increases in risk acceptance that routinely follow from the sense of security that safety innovations provide.[33] In each instance, perceptions of safety end up generating risky behavior that can potentially erode any gains that were made and may even lead to a greater number of accidents. As with moral hazard, the underlying concern is that efforts to manage risks may be counterproductive if they encourage greater risk acceptance and facilitate the displacement of that risk onto others.

Are Drones a Moral Hazard?

Kreps and Kaag argue that drones create a moral hazard by making it possible for their users to wage wars without experiencing the full costs of fighting. Drone warfare shifts the burdens of war onto others, first, the authors maintain, because drones are particularly dangerous for civilians, and second because pilot invulnerability makes it virtually impossible for their users to experience the human costs that are typically associated with armed conflicts. Kreps and Kaag go on to argue that the transference of risks makes risky behavior more attractive, allowing drone users to increase the scope and frequency of their military operations with deadly consequences that are not felt domestically. They cite the American targeted killing program as an example of this. Because the American military can launch drone strikes in foreign countries with impunity, it is drawn into a false sense that this violence is effective for countering terrorism:

> Killings and signature strikes have always been in the repertoire of military planners, but never, in the history of warfare, have they cost so very little to use. Historically, there have been significant risks associated with these types of military activities. First, a nation-state had to risk its soldiers or operatives. And if it risked the lives of soldiers, it risked public censure if these individuals were killed or captured. And if such a strike was successful and publicly endorsed, there was still the risk of retribution. Drones have allowed some governments to obviate most, if not all, of these costly risks. As US military forces acquire more autonomous drones and hire private security firms to operate them, these killings will become even less expensive for the American citizenry.[34]

To some extent, the moral hazard that Kreps and Kaag identify is contained within the government, affecting those who are directly responsible for military affairs. They maintain that policymakers and members of the military are susceptible to the moral hazard because they lack the moral sensitivity needed to understand the problem and are compelled by the benefits of drone warfare to underestimate the magnitude of the risks being borne by others. Nevertheless, because citizens in a democracy are in a position to influence decisions to use force, the moral hazard affects the American public as well. The civilian public may take on a false sense of confidence in the country's ability to wage wars without bearing the costs of fighting, which may in turn make people more willing to initiate wars that impose high risks on foreign populations and more willing to wage wars that have dubious prospects of success simply because they are relatively risk-free.

More general findings about how states and political leaders respond to risks provide grounds for thinking that the moral hazard is genuine. Kahn describes a "paradox of riskless warfare" that emerges when technological advancements or political decisions result in radical asymmetries of risk between opposing belligerents. He contends that war can only truly qualify as war when there is "reciprocal imposition of risk" between combatants on opposing sides, and that "[w]ithout the imposition of mutual risk, warfare is not war at all."[35] This is paradoxical because Kahn recognizes that there are powerful incentives driving states to pursue advantages over their opponents, even as these could potentially become radical risk asymmetries that violate the logic of war. This goes beyond Kreps and Kaag's concern about drones making it easier to wage wars to present an existential concern about sustaining the concept of war itself.

Many other commentators invoke similar arguments against drones, though often without explicitly mentioning moral hazard or the Peltzman effect by name. It is worth considering a few examples of this to appreciate how pervasive this reasoning is and to see the subtle variations in the mechanisms that are described. It is also revealing that this argument is largely speculative, resting on the assumption that the logic of moral hazard can be generalized from other fields of research without empirical findings specific to drones.

Chamayou follows Kreps and Kaag in arguing that greater reliance on drones could make it easier to launch wars and more difficult for civilians at home to perceive the effects of violence. "Freed from the constraints imposed by reciprocal relations, will the drone masters be able to continue to demonstrate virtue and to resist the temptation to commit injustice with virtually no sanction imposed?"[36] This way of phrasing the problem emphasizes the central question of whether those using the drones will, in the absence of costs imposed by opposing belligerents, be able to restrain themselves. This highlights the salience of whether public opinion will be sensitive to moral, strategic, or other considerations that will urge restraint even when the risk of sustaining casualties is low or when peaceful alternatives are available. Chamayou's reasoning is plausible based on other studies of moral hazard, but he does not test whether this problem applies to drones.

Assessing the effects of drones on perceptions of risk also addresses broader concerns in research on risk management in military affairs. Shaw contends that the United States and other advanced militaries are responsible for a phenomenon he calls "risk transfer militarism," which is characterized by a style of fighting that shifts danger away from soldiers and onto civilians, enemy combatants, and local allies.[37] The decision to transfer risks is evident in the reliance on bombing and other types of violence that reduce soldiers' vulnerability, such as drones. These weapons not only prevent the American military from endangering soldiers in combat but also lead it to rely more heavily on allies who can provide ground forces and whose soldiers are therefore open to being injured or killed. Seen from this perspective, the moral hazard presented by drones may be a manifestation of a broader strategic and political decision about how the United States seeks to distribute the risks of war and, in doing so, generates the new danger of losing contact with the horrors of violence.

Similar arguments inform a large body of research attempting to come to terms with what it means to fight wars when the soldiers and civilians are insulated to such an extent that real war transforms into a "simulation,"[38] "virtual war,"[39] "virtuous war,"[40] "war without warriors,"[41] or a "spectator sport."[42] These claims commonly point to drones as one of the primary culprits in transforming attitudes toward violence. Because these weapons platforms offer both pilot invulnerability and target selectivity, they are described as the logical outcome of American or Western strategic preferences. By mediating the experience of war and only showing the effects of bombs and missiles through a computer screen, drones raise still more concerns of cultivating the kind of detached sense of risk that is blamed for making war too easy. Here the concern is not with war itself but rather with a particular kind of war: unnecessary war that would not take place were it not for the public's distorted perceptions. If this type of war only happens because it is easy, rather than because it is necessary for national self-defense, then there must be peaceful alternatives that would become apparent by shattering misperceptions.

Some wars may be largely unavoidable. A country that is invaded by a hostile foreign state has little choice but to take up arms in self-defense. There is, therefore, little likelihood that moral hazard or related disruptions to risk assessment will cause defensive wars. If moral hazard is a genuine concern, its effects should be clearest when it comes to what are sometimes called "wars of choice." These are wars that states initiate even when there are alternatives to fighting. As Haas explains, "wars of choice tend to involve stakes or interests that are less clearly vital, along with the existence of viable alternative policies, be they diplomacy, inaction, or something else but still other than the use of military force."[43] Moral hazard is most serious in these cases, where a war might not have happened if not for the distortions in risk assessment brought about by the capabilities of drones. This leads us to the second expectation we explore in this chapter: support for an attack that takes place despite the availability of peaceful alternatives will be higher when drones are involved.

Drones and Moral Hazard

This chapter's first expectation holds that the technological capabilities of drones will lead individuals to believe that their battlefield objectives are more likely to be met with drones than with other types of military force. To assess this expectation, we analyze responses to a question included in the survey experiment with a representative sample of American adults introduced in detail in chapter 3. Recall that this experiment randomly assigned respondents to read a news story describing a planned attack on terrorist compounds. The respondents read stories that described the attack as being conducted by drones, air strikes, or ground troops. They then answered questions measuring their responses to the news article. One of these items asked respondents to indicate how likely the attack would be to achieve its military objectives. Responses fell on a seven-point scale ranging from "extremely likely" to "extremely unlikely" and are summarized in figure 3.4. Our first expectation in this chapter is that the percentage of respondents expecting the attack to have a higher chance of success would be considerably greater for those who read about drone strikes than for those who read about an air strike or a ground invasion. This is clearly not the case. The distribution of responses is remarkably similar across the three conditions, and a statistical test indicates that there are not significant differences among the three groups of respondents. This suggests that, on average, the public does not attribute clear or obvious advantages to drones over other types of military force.[44]

To gain more traction on this issue, we also conducted a survey experiment designed to assess the degree to which individuals fell victim to moral hazard when considering the utility of drone strikes, using respondents recruited from Amazon's Mechanical Turk.[45] Our experiment builds on work by Aaron Hoffman and his colleagues. They show that existing experimental and public opinion studies of military success systematically overstate support for the use of force. The reason is that these works typically do not discuss *alternatives* to American military action. Experiments follow a pattern of presenting respondents with a range of military options to choose from but do not provide information about what nonmilitary options are available or how likely these are to succeed. The public, they suggest, generally assumes that military operations are pursued when nonviolent methods of conflict resolution have failed. As Hoffman and his colleagues put it, this type of experiment prompts participants to believe that "force is a last resort—to be used only after other feasible alternatives are exhausted." Omitting diplomatic alternatives signals that they are unavailable and that some kind of military response is required.

They conduct a series of studies demonstrating that when respondents are informed about the existence of viable alternatives to the use of force, support for military action declines.[46] Conversely, when the information about alternatives is not mentioned or when the prospects of diplomatic success are low, the support for war is similar to that found in experimental designs that only include military options. This is evidence that "[o]mitting information about the feasibility of diplomatic strategies can induce people to conclude that non-war courses of action are unattractive, strengthening their willingness to back military action." The potential human costs play a central role in the causal process the authors outline. They find that raising the expected casualties dampens enthusiasm for attacking, such that the alternatives to war look more attractive as the projected casualties increase. Lowering the number of expected casualties makes military options more appealing even if peaceful alternatives are mentioned. By this account, casualty sensitivity is best evaluated in terms of how much loss the public will tolerate when the available options, their likelihood of success, and the policy objective are also factored in.

Military action is viewed as a last resort because it is costly and risky. The preference for nonmilitary strategies when these are available, or when high casualties are anticipated, is a natural consequence of risk aversion and the associated moral abhorrence for causing needless harm. As we saw in chapter 3, a key cost for many individuals is the possibility of military casualties. Drones, of course, permit the United States to use force without risk to soldiers. If the logic of moral hazard outlined above reflects how people think about the risks and costs of conflict, the findings reported in the Hoffman experiment should disappear when drone strikes are available. Since drones reduce the expected military losses to zero, the logic of moral hazard indicates that the public should be willing to support their use even when the chance of success is low and if there are viable alternatives to military force. The public should feel the lure of the easy "wars of choice." On the other hand, the moral hazard reasoning would be contradicted if the public continues to favor nonmilitary options even when drones can be used to prevent American losses.

Our experiment in this chapter tests this expectation. Like other experiments reported in this book, it takes the form of a news story describing a planned attack. This time, respondents read about a plan for American forces to support Iraqi troops defending a strategically important bridge from militants. It had two treatment arms. The first, similar to the experiments reported in chapter 3, varies the attack type, which was described as being carried out by a drone, aircraft with onboard crews, or ground troops. The second arm, building on the work of Hoffman, adds information about alternatives to military action. Half of the respondents read that the secretary of defense, responsible for authorizing the strike, "was overheard saying that this decision was an easy one because the alternatives to American intervention . . . such as dropping weapons to the Iraqi troops or airlifting more Iraqi troops to defend the bridge, were unlikely to work." The other half of the respondents read a nearly identical text, except they were informed that the secretary said the decision was a "difficult" one because the alternatives were "likely" to work. This replicates the way Hoffman and his colleagues presented their experiments, making our results comparable to theirs. Framing the experiment in this way produces six conditions: a drone strike with good or with bad alternatives to the use of force, an airstrike with good or bad alternatives, and an attack by ground troops with good or bad alternatives.

As in our previously reported experiments, the dependent variable is support for the attack described in the news article; respondents can strongly oppose, somewhat oppose, somewhat support, or strongly support this use of force. Our key comparison is across respondents that are assigned to treatments with the same attack type but different information about the alternatives to military force. The moral hazard theory would predict that among respondents who read about attacks by ground troops, support would be higher among those who were told there were only bad alternatives to the attack than among those who were told that good alternatives existed. Respondents in the later condition, wishing to avoid the risk of military casualties and knowing that there are other ways of achieving the same objective, should be less enthusiastic about using force. The moral hazard theory would expect a quite different pattern across respondents who read about a drone strike but differed in what they were told about the alternatives. These respondents should, according to the logic of moral hazard, be indifferent to the availability of alternatives, since drones are a low-cost means of achieving the objective. This expectation generates weaker predictions about respondents assigned to the air strike treatments. Since the pilots of strike aircraft are presumably at less risk of harm than ground troops, a plausible extension of the logic of moral hazard might hold that the effect of good alternatives would be weaker than in the case of ground troops but stronger than for drone strikes.

To assess these predictions, we first divided the sample of respondents into three groups: those who read about drone strikes, air strikes, and the use of ground forces. We then created a dichotomous measure of assignment to a condition describing the alternatives to military action as good or bad. For each group of respondents, we regressed this dummy variable on our dependent variable of support for the use of force using ordered logistic regressions and including standard control variables: gender, education, party identification, age, and minority status. The results are presented in table 5.2.

The predictions of the moral hazard argument were not borne out by our experiment. For each attack type, the key independent variable—the existence of good alternatives to military action—is negative and statistically significant. This means that respondents who were informed that there were good alternatives expressed less support for military action, regardless of the type of attack. This is to be expected among the group of respondents who read about an attack by ground forces. But the fact that the same relationship holds among those who read about a drone strike is

TABLE 5.2. Regression Results for Attacks with Good and Bad Alternatives

	Drone Strike	Air Strike	Ground Troops
Good alternatives	-.666*	-.462*	-.791*
	(.200)	(.193)	(.199)
Gender	-.526*	-.845*	-.133
	(.210)	(.201)	(.201)
Education	1.101*	.0418*	-1.405*
	(.564)	(.193)	(.562)
Party identification	2.106*	1.390*	1.292*
	(.382)	(.374)	(.396)
Age	1.212*	.749	-.434
	(.680)	(.703)	(.700)
Minority	.201	.032	.137
	(.239)	(.214)	(.227)
Likelihood ratio	842	915	872
AIC	860	933	889
Observations	381	387	379

Note: Standard errors in parentheses; * $p < .05$.

inconsistent with the prediction that the availability of drones will increase the public's willingness to initiate "riskless wars."

The results in table 5.2 only tell us if the effect of the existence of good alternatives is statistically significant or not. It is possible that the size of the effects of this variable are quite different across the three groups. For example, it might be that such alternatives reduce support for drone strikes, but by a much smaller amount than for ground troops. To investigate this possibility, figure 5.2 plots the changes in the predicted probabilities for each level of the dependent variable across respondents assigned to conditions with good and bad alternatives. We see that the substantive effect of this variable is similar across the three attack types; respondents who read about good alternatives to military force were between 10 and 20 percent less likely to strongly support or to support the planned attack. These differences were largest among respondents who read about the use of ground troops, but were still of similar magnitude to those for whom the attack was described as a drone strike.

The experiment included one additional question that addresses a potential alternative explanation. As with the representative sample of respondents from the experiment reported in chapter 4 and figure 5.1, in this experiment we also asked respondents to report whether they expected the attack described in the condition they read to succeed. The percentages of respondents reporting each possible answer choice is reported in figure 5.2.[47] Among the respondents in this experiment there was very little difference in the expectations that the attack would succeed across the three attack types. In particular, note that the two conditions involving drone strikes do not produce substantially higher expectations of success than do air strikes or the use of ground troops. This offers further evidence against expectation 5.1 by showing that respondents did not think that drones were more or less likely to secure victory than other methods of attack.

Conclusion

Our finding that the public does not see drones as being any more or less effective than manned aircraft and ground forces indicates that these different forms of military force are viewed as interchangeable when it comes to their ability to achieve objectives. If anticipated success is the decisive factor in determining support for war, then a war waged with drones will not appear significantly more attractive than one conducted by

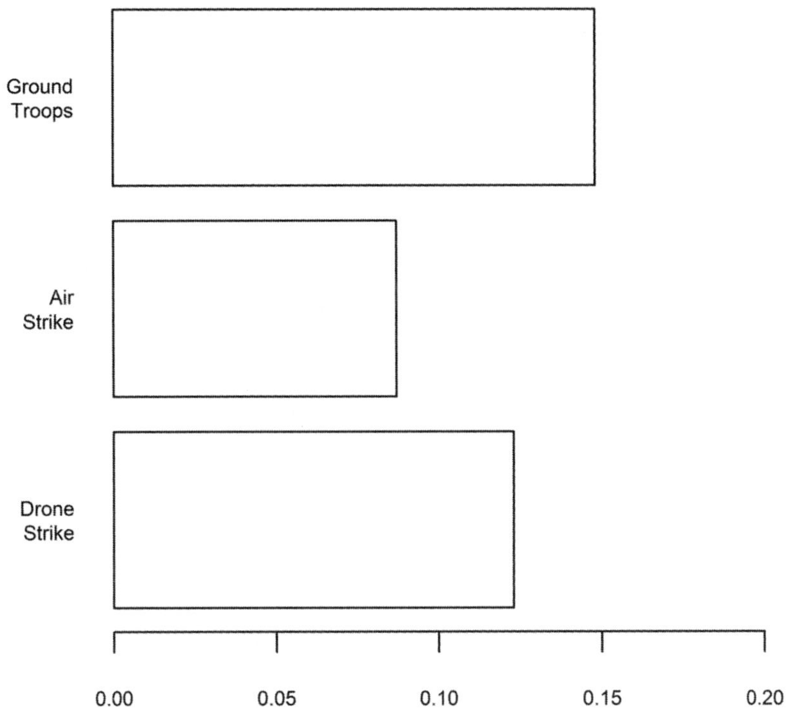

Fig. 5.1. Differences in predicted probabilities of supporting attack with good and bad options

other means. Any difference in enthusiasm for drone attacks compared to alternatives would have to come from something other than shifts in the chance of victory. In chapter 3, we suggested that the most important reason that drones increase support for the use of force is they render American military personnel invulnerable from attack. The findings in this chapter, that drones are not viewed as more likely to achieve their objectives than are other types of attacks, lends further credence to this conclusion. This should help to allay some of the concerns that drones may lower the threshold for war. It means that drones are not likely to promote greater confidence in military force as a way of reaching foreign policy goals. The military option remains a singular approach that is apt to be judged by the same standards about likelihood of success as in previous conflicts when drones were not available.

These conclusions have some implications for the work on elite fram-

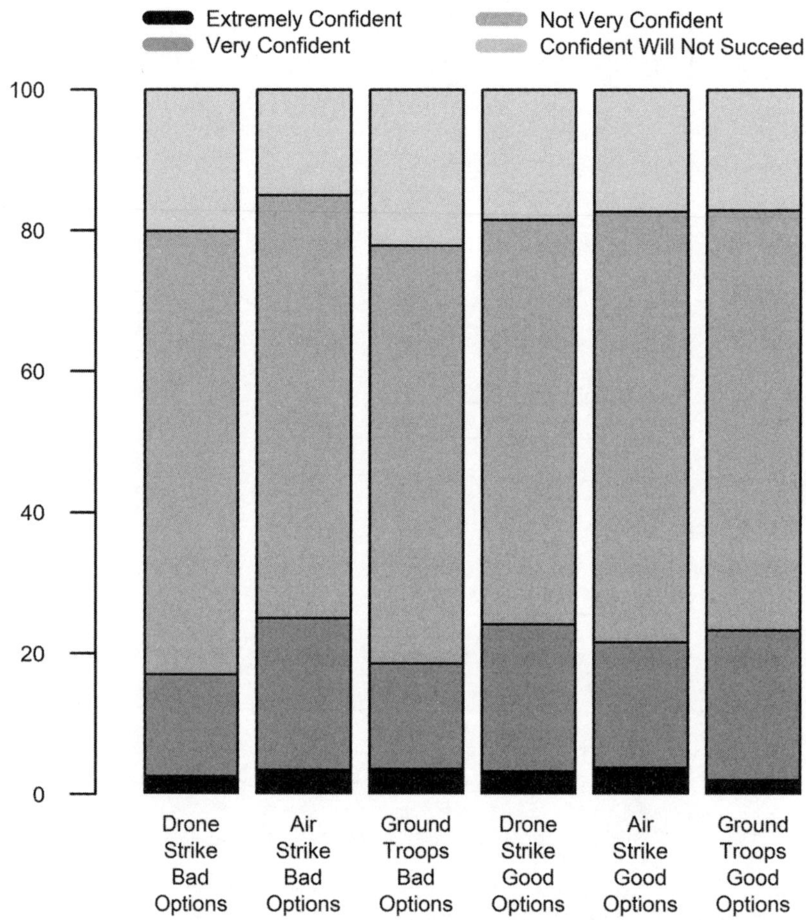

Fig. 5.2. Expectations of success by mission type and good and bad options

ing we have discussed in previous chapters. Elites attempting to mobilize support for a drone war may have limited success if they present drones as being more effective weapons platforms for achieving military objectives. Instead, they are likely to produce greater improvements in public opinion when they focus on the protection of American soldiers. Opponents of drone strikes may likewise achieve little if they characterize these machines as less effective than alternative forms of military force and would be better served by emphasizing their impact on civilians, which we discuss in the next chapter. The murkiness of concepts like victory and defeat in the

War on Terror that has come to define drone operations only adds to the challenge for elites on either side who attempt to promote drones as more or less effective than traditional weapons and tactics.

Anticipations of the likelihood of success get at several important normative considerations when it comes to determining when wars are justified and how they should be fought. First, likelihood of success is often taken to be one of the core principles of just war theory.[48] A futile war, the reasoning goes, is immoral because it inflicts suffering without any chance of producing redeeming benefits. Predictions of success may therefore inform judgments about whether a war is justified. With drones having roughly the same degree of perceived efficacy as manned aircraft and ground forces, they are unlikely to influence this dimension of the moral and legal reasoning that goes into determining when wars are justified. Drones may alter the calculations people make about when to fight, but this result helps us focus on how they do this by eliminating some of the possible explanations. We can see that although the likelihood of victory is normatively important, perceptions about the morality of drone warfare are more heavily shaped by other influences.

Second, the likelihood of winning may influence how belligerents choose to fight. Alexander Downes argues that states are more inclined to target civilians when they are desperate to win—a condition that occurs when the costs of losing are high and the chances of winning seem to be remote. "According to the desperation logic, states that are embroiled in costly and prolonged struggles become increasingly desperate to snatch victory from the jaws of defeat and reduce their own losses."[49] As this quotation indicates, the anticipated number of casualties interacts with anticipated success just as in public opinion research, making it essential to distinguish these factors in experiments. But he adds the additional worry that changes in opinion might translate into shifts in the conduct of wars, with terrible results for the civilian victims. Drones could be advantageous or disadvantageous depending on whether their availability changes the conditions under which states experience the desperation that Downes describes and become willing to deviate from the norms of fighting. Based on our results, their impact appears to be neutral and therefore unlikely to influence the civilian public's desperation to win.

The results of our experiment in this chapter show that effects of moral hazard or the Petlzman effect do not appear to hold when it comes to combat drone technology. The ability to use drones does not increase the

American public's willingness to engage in risky military ventures when nonviolent alternatives are available. Rather, good alternatives to war are consistently preferred regardless of what weapons are available, and violence only becomes more attractive as the alternatives disappear or become less promising. We should therefore be skeptical about arguments that the advent of drones will radically alter the conditions under which the United States decides to initiate wars. As we have shown in previous chapters, drones help to build support for war, but their effect tends to be modest and short of what the United States would need to initiate wars with impunity.

There is an important caveat to this conclusion about moral hazard. We have focused on one manifestation of moral hazard linked to drones, the one most commonly expressed in the literature and potentially the most concerning because it raises the possibility of an increase in the prevalence of armed conflicts. Nevertheless, there are other variations of the moral hazard argument that cannot be directly tested using public opinion data. For example, Chamayou argues that "[i]f the military withdraws from the battlefield, enemy violence will turn against targets that are easier to reach. Even if the soldiers are beyond reach, civilians are not."[50] Killmister likewise contends that relying more heavily on drones is morally questionable because this will induce enemies to pursue terrorism. These arguments rely on the moral hazard logic insofar as they suggest that the displacement of risk, in this case from soldiers onto civilians, may encourage states that use drones to become more risk-acceptant. Future research should explore these and other ways drones may upset risk calculations during wars. However, the analysis will have to focus less on attitudes within the country using drones and more about how these weapons platforms change strategic calculations for those who come under attack.

CHAPTER 6

Drones and Sensitivity to Civilian Harm

We saw in chapters 2 and 3 that individuals are, on average, more willing to support the use of force when attacks are carried out by drones rather than by ground troops, and that the invulnerability these weapons platforms create for American military personnel is an important part of this relationship. This creates an ethical dilemma: by lowering one cost of war—military casualties—drones may make it politically easier to start or intensify wars. These wars could result in civilian deaths. As discussed in the introductory chapter, a major concern of critics of the use of armed drones is that they may lead to the substitution of military casualties with civilian casualties. There is also the related fear that these civilian casualties would be ignored if drones promote public disengagement from the overall costs of fighting. In this chapter, we suggest that this way of thinking about drones and civilian casualties overlooks how this technology changes public expectations about the likelihood of civilian casualties.

There is some evidence that the American public is not overly concerned about civilian deaths during war. As we discuss in the next section, some systematic studies of the issue find that the deaths of noncombatants have little effect on American public opinion. Other studies, though, present evidence that Americans are quite concerned about the possibility of civilian casualties. Our conjoint survey experiment in chapter 2 found that increased expectations of civilian casualties were associated with sizeable reductions in support for attacks. This is evidence of such a high concern for civilians that expected civilian casualties can constrain US military options. What can account for these contradictory findings? We contend that one answer can be found by looking at how the availability of drones changes the perceived salience of civilian casualties.

Drawing on insights from psychology about "counterfactual thinking," we theorize that the selectivity of drones—that is, their ability to distinguish legitimate targets of military action from noncombatants—raises expectations about the results that will obtain when they are used. Drone technology may make it easier to avoid civilian casualties, and when prompted with this information, the public adjusts its expectations. Other precision weapons may have a similar sensitizing effect, but drones deserve special attention because they have been the most heavily publicized tool for launching selective attacks throughout the War on Terror. Moreover, the US military's overall preoccupation with achieving greater precision across a range of different weapons systems fuels expectations that drones will contribute to the overall reduction of civilian losses.

A risk of civilian casualties should reduce support for the use of force when drones are available, and attacks with drones that result in civilian casualties should heighten concern about the casualties that are inflicted compared to similar attacks using less selective weapons or attacks that place military personnel at risk. This process of counterfactual reasoning drives people to pay more attention to civilian harm because such outcomes conflict with their expectations. It raises the salience and prominence of civilian casualties in the minds of the public and leads to greater dissatisfaction and regret when drone strikes produce civilian deaths.

A first experiment tests the chapter's first two expectations (summarized in table 6.1) derived from the logic of counterfactual thinking about how the selectivity of drones makes individuals more sensitive to civilian harm. Respondents primed to expect fewer civilian casualties expressed more regret, more sympathy with victims' families, and less satisfaction than did those primed with a higher risk of civilian deaths, even though the actual outcomes across treatments were identical. This effect weakens when precision strikes are carried out by manned platforms.

TABLE 6.1. Expectations

Expectation 6.1: Compared to otherwise identical attacks with non-precision weapons, attacks with drones that result in civilian casualties produce less satisfaction, more regret, and more sympathy for the victims.

Expectation 6.2: The effects of civilian casualties caused by attacks with precision weapons on satisfaction, regret, and sympathy will be attenuated when these weapons place military personnel at risk of harm.

Expectation 6:3: Support for the use of force declines when the authorities forgo opportunities offered by precision weapons to reduce the chance that the use of force will result in civilian harm.

An important counterargument is that many view civilian casualties as tragic but unavoidable accidents that result when military force is employed. We further argue that the development and use of drones should alter this perception of the origins of civilian harm. Most people want to avoid civilian casualties, and they expect political and military leaders to take reasonable steps to achieve this goal. But these steps can conflict with the pursuit of other military objectives; for example, deciding not to attack a potential target because doing so would result in civilian harm might expose American troops to greater danger or leave an insurgent group free to engage in more violence in the future. These conflicts recede, though, when drones are employed. Their technological characteristics enable force to be used in ways that more carefully distinguishes military from civilian targets and that has the potential to reduce the risk of civilian harm.

We argue that people expect their leaders to exploit these advantages and to plan their attacks in ways that reduce the risk of killing civilians, especially when doing so does not reduce their ability to achieve military objectives. A second survey experiment assesses this third expectation and finds that support for the use of force declines when planned attacks unnecessarily endanger civilians. This effect strengthens when respondents are informed that the attack violates international law, which suggests that legal and moral norms exert an independent influence on assessments of the use of force. Importantly, the differences remain even when the targeted group targeted is described as directly threatening the United States, indicating that individuals are not willing to "trade" risking civilian death for greater security.

Counterfactual Thinking about Civilian Casualties

What effect might the deaths of foreign civilians in drone strikes have on the American public's support for the use of force? Existing research does not provide the straightforward answers that work on military casualties does. Much less attention has been devoted to analyzing if and how civilian deaths influence attitudes toward war. Some systematic studies of the issue find that the deaths of noncombatants exercise little effect on American public opinion. John Mueller's review of aggregate polling data from the Second World War through the 1990s leads him to conclude that the American public is not much influenced by such casualties. John Tirman's historical analysis reaches a similar conclusion. One remarkable finding is that civilian deaths at the hands of American forces in some conflicts, especially the wars in Korea and Vietnam, were far greater than those resulting from drone strikes, but they seemed to exert little influence over public attitudes. On its face, this would seem to indicate that even very high levels of harm to foreign noncombatants has little effect on support for the use of force.[1]

Nevertheless, there is strong evidence in favor of the opposing perspective. Some studies show that civilian casualties might influence public attitudes under certain conditions. Richard Eichenberg found that survey questions mentioning civilian casualties led to a reduction in support for the use of force. The size of the effect of civilian casualties was similar to that of military casualties and as large as many of the principal policy objectives, such as foreign policy restraint and humanitarian intervention.[2] Experimental work by Johns and Davies finds that information about civilian casualties reduces support for the use of force.[3] Other studies indicate that concern about civilian casualties has increased in recent decades.[4] This helps to reconcile the opposing viewpoints by introducing the possibility that opinions may vary over time, yet the causal processes responsible for the growing concern remain unclear.

Existing research, then, has reached conflicting conclusions about the effects of civilian casualties, with some finding that it has no effect on support for the use of force and others concluding that support declines when civilians are harmed. We suggest that one reason for these different findings is that the public is, often implicitly, balancing the costs of military casualties against the costs of civilian harm. In their important work on support for the use of force, Gelpi, Feaver, and Reifler say that "the American public views civilian casualties much the same way they view military casualties—as a necessary evil to be minimized, but tolerable under the right conditions." But any trade-off between minimizing risks to military personnel and to noncombatants disappears when drones are employed. Drone strikes place no American service members at direct risk of harm. How might this influence support for the use of force? One perspective might hold that the weakening of the trade-off between civilian and military casualties should unambiguously lead to more support for military action. When this trade-off is sharp, many individuals will be willing to "exchange" a higher risk of civilian harm for a reduction in the risks to military personnel. But drones render this mental exchange unnecessary;

it is now possible for people who opposed an attack because of the risk of military casualties to support the attack.

As we discussed in chapter 3, much of the research on drones takes this perspective by speculating that the absence of military casualties will lead to a corresponding willingness to wage more wars. The central argument in this chapter is that heightened concern about noncombatant casualties is a direct consequence of a weapon's selectivity. Drones, for example, are perceived by many as selective in their ability to identify and destroy a target while minimizing harm to nearby people and structures. We argue that this selectivity influences not only support for the use of force, but that it also raises expectations about the outcome of military action: as military technology becomes more selective, and this selectivity is communicated to individuals, they become less tolerant of paying costs that might have been avoided. As Sapolsky and Shapiro noted some time ago, there are "ratchets in our war experience"; the successful use of precision weaponry in a conflict today creates the expectation that the same degree of success will be achieved in tomorrow's wars.[5] The US military has shown a strong desire to employ precision weapons throughout the post–Cold War era, with drones emerging out of this quest for precision and becoming its most prominent expression.

Insights from the study of "counterfactual thinking" explain this ratchet effect. The key work is by Kahneman and Tversky, who introduced the idea that counterfactuals can be integrated with the study of judgment and decision making.[6] In this tradition, counterfactuals are conditional propositions with an antecedent and a consequence, where the antecedent is the action or decision of an individual or group and the consequence is the outcome that results from this action. Counterfactual thinking involves a mental modification to the antecedent, and then imagining how this modification would alter the consequence. Upward counterfactuals, which are most relevant here, occur when the outcome is judged to be inferior to what would have obtained if the imagined alternative to the antecedent had been chosen.[7] An example of an upward counterfactual is "if the drone strike had not been authorized, the civilians would not have been killed." Upward counterfactuals have been shown to influence emotions and judgments. For example, they drive feelings of regret, defined as the recognition that a different antecedent action would have resulted in a superior outcome.[8] Respondents are willing to provide more compensation to victims of crime and negligence when an upward counterfactual is mentally available to them.[9] These effects are the consequence of two mechanisms: contrast effects and causal inference effects.

Contrast effects arise "when a judgment is made more extreme via the juxtaposition of some anchor or standard."[10] This is nicely illustrated in a study of Olympic athletes, which found that those winning silver medals were less satisfied than those awarded bronze metals. Silver medal winners generated upward counterfactuals, comparing their outcome with winning a gold medal, while bronze medal winners thought in terms of downward counterfactuals, contrasting their medal with the outcome of no medal.[11] The reference value that an individual uses to evaluate an outcome, then, has a decisive influence on the effect and judgments produced by counterfactual thinking.[12]

Causal inference effects occur when the imagined counterfactual is identical to the facts except for the action in the antecedent. This means that in the mind of the individual, the antecedent action is the cause of the outcome. The practice of contrasting an otherwise identical counterfactual with the actual decisions and events and outcome highlights the causal link between the action and the outcome. Causal inferences are more likely when the antecedent is more "mutable" or subject to change. It is more difficult to imagine changing immutable causes, such as gravity. Actions and decisions, in contrast, are often perceived as more mutable, since in many cases it is possible to imagine that a different action had been chosen. Based on this reasoning, our first expectation is that compared to otherwise identical attacks with nonprecision weapons, attacks with precision weapons that result in civilian casualties produce less satisfaction, more regret, and more sympathy with the victims.

The contrast effect and causal inference effect contribute to this difference. Drones' selectivity and pilot invulnerability allow political leaders to argue that the weapons are less likely to produce negative unanticipated effects. The political and military leaders that develop and deploy precision weapons justify them, in part, because they are more likely to produce desirable outcomes (such as killing the enemy) and less likely to result in undesirable consequences (such as killing civilians). The reference value for civilian casualties that individuals adopt for drones, then, should be lower than that for other, less selective weapons. If an attack does inflict civilian casualties, the contrast between the reference value and the outcome will be greater when drones are employed, leading to stronger affective and cognitive responses.

A second relevant feature of drones is their tight integration of intelligence collection with targeting and attack decisions. This integration is one of the factors that makes these weapons so precise and contributes to the contrast effect. It is also an advantage over some other types of precision weapons. As we pointed out earlier, the decision to arm drones was initially taken to avoid the lag time between identifying a target and the impact of missiles. Drones increase selectivity beyond what some other precision weapons can achieve by reducing this time from the minutes it takes a cruise missile to arrive at its destination—during which a civilian may wander into the kill zone—to a few seconds.

Combining reconnaissance and attack roles makes the decision to launch an attack more mutable. Decision makers, including military personnel piloting the weapon or political and military leaders who authorize its use, have more, higher quality, and more up-to-date information about the target. This is particularly clear from the American military's drone strikes against suspected terrorists, which are often preceded by hours or even days of monitoring the target to confirm that the person is a combatant and to select the right moment of attack. This means that it is easier to imagine that there might have been intelligence that would have led to calling off an attack that results in civilian harm.

Contrasting an old-fashioned "gravity" bomb with a guided missile fired from a drone illustrates this effect. It is not difficult to imagine many factors, some of which are not particularly mutable, preventing a gravity bomb from hitting its target: wind, the inability to view the target due to weather, or human errors in making the complex calculations needed to accurately drop the bomb. In the case of an attack from a precision weapons platform such as a drone, imagining a different and better outcome is considerably easier. The weapon controllers are able to monitor the target in real time, can rely on software the guide the missile to a specific point, and can call off an attack if a civilian wanders into the target area. All precision weapons improve on gravity bombs when it comes to reducing the risk of civilian casualties, so drones are not unique in this respect. Drones should be seen as one part of a broader category of precision weapons systems that aim to reduce the risk to civilians. Drones offer some advantages over other technologies in this group by integrating functions and reducing lag time between target identification and a weapon's impact, but all precision weapons systems may have some influence on altering the anticipated number of civilian casualties that would result from a strike.

The important work by Gelpi, Feaver, and Reifler included a survey experiment that compared the effects of military and civilian casualties in a hypothetical war with North Korea. Civilian casualties reduced support for the war about as much as did US military casualties. Furthermore, a majority agreed that military planners should limit harm to foreign civilians even if this placed American military personnel in greater danger. Most importantly for our purposes, their survey experiment directly asked respondents to rate the importance of limiting American military casualties compared to civilian casualties. Most respondents stated that the former goal was more important.[13] This suggests a second expectation: the effects of civilian casualties caused by drones on satisfaction, regret, and sympathy will be attenuated when these weapons place military personnel at risk of harm. The logic of this expectation is that respondents would prefer to avoid both civilian and military casualties. When it is impossible to guarantee this outcome, the possibility of military casualties makes them somewhat more willing to tolerate civilian casualties.

In sum, individuals' assessments of an attack with drones that may cause harm to civilians are shaped by the difference between the reference point individuals adopt and the outcome. Negative differences lead to more negative assessments. This means that assessments of an attack with the same outcome can differ across individuals who begin with different reference points. These reference points, in turn, are influenced by how precise a weapon is depicted by opinion leaders such as the media. Finally, this effect is blunted when military personnel are placed at risk.

Counterfactual Thinking and Selectivity: A Survey Experiment

We conducted a survey experiment in the summer of 2013 to better understand how counterfactual thinking influences attitudes about drone strikes and civilian casualties. After collecting information about their demographic characteristics, political attitudes, and attention to current events, respondents were randomly assigned to read one of four hypothetical news stories (the treatments, survey administration details, demographic statistics, and regression results can be found in the appendix to this chapter). All four treatments described a planned attack by the United States military on the hideouts of insurgent groups in Pakistan that had attacked American troops in Afghanistan. These hideouts were described as being near civilian residences. The vignettes stated that civilian casualties were one possible outcome of the planned attack.

The treatments varied the military technology that would be used in the attack and the assessment of military experts regarding the likelihood of civilian casualties. In the "high selectivity, unmanned" treatment, the attack was described as being carried out by a drone equipped with video surveillance that allowed the attacks to be launched when civilians were not nearby and using highly accurate missiles that would minimize the risk of civilian casualties. The vignette also stated that in 2012, only 3 percent of those killed in similar drone strikes were civilians. This number is drawn from the estimates of the Bureau of Investigative Journalism, which carefully tracks drone strikes and civilian casualties.[14] This information was intended to provide respondents with a low reference value regarding the likelihood of civilian harm.

The "high selectivity, manned" attack was identical to this first treatment but described the attacks as conducted from piloted aircraft. This treatment stated explicitly that the insurgents lacked weapons capable of firing on the attacking aircraft, meaning that the crews were not at risk of harm from the enemy. Differences between these two treatments allow an assessment of the expectation that the effects of civilian casualties should be attenuated when the attack is conducted with manned rather than unmanned platforms. This design creates a difficult test for this expectation, since respondents are explicitly told that the air crews cannot be attacked. Respondents must imagine some way the crews could be harmed, for example equipment malfunctions over hostile territory or the possibility that the insurgents have acquired weapons unknown to the United States.

The "moderate selectivity" treatment described the attack as bombs dropped from aircraft. These aircraft would fly at high altitudes so that their crews would not be at risk from insurgent fire and would drop bombs on the insurgents' hideouts. These bomb attacks were described as possibly producing civilian casualties because of their larger blast size and the fact that the pilots could not easily determine if civilians were near the targets. In this treatment, experts stated that civilian casualties were "possible" and that 35 percent of those killed in similar attacks in 2012 were classified as civilians. Since the goal of this treatment was to prime respondents to imagine a much higher likelihood of civilian harm than those assigned to the high-precision treatment, we selected a number here that was an order of magnitude larger than in the high-precision treatment. Finally, the "low selectivity" treatment was identical to the moderate selectivity treatment except it stated that 50 percent of those killed in similar attacks were civilians. The third and fourth treatments assess the expectation that differences in respondents' degree of sympathy and regret for civilian casualties should be lower than for the unmanned, high-selectivity treatment.

Immediately after reading the vignette, respondents in all four treatment conditions were asked if they supported the plan to attack insurgent bases, their assessment of the likelihood of civilian casualties, and their level of satisfaction with the attack plan.[15] All respondents were then informed that the attack had been carried out and had resulted in the death of the targeted insurgents as well as civilians. After learning this information, they were again asked questions regarding their level of satisfaction with the attack plan. Respondents also indicated how surprised they were that the attack resulted in civilian casualties and how important they thought it was that the United States issue an apology for killing civilians. They were also told that in past conflicts, the United States has offered compensation to victims' families, and they were asked how much compensation should be offered in this case. Possible responses ranged from no compensation to $40,000 or more in increments of $5,000. Respondents were informed that the average amount suggested was $20,000.[16]

The theory of counterfactual thinking creates three expectations about how responses should differ across treatments, even though the outcome was identical for all respondents. First, those primed by the unmanned, high-selectivity treatment to expect fewer civilian casualties should be more sympathetic to the victims and favor higher levels of compensation to victims' families. This measure draws from an important study that assessed how counterfactual thoughts accounted for sympathy with victims of violent crime.[17] Second, those assigned to the high-selectivity, unmanned treatment should be more willing to issue an apology, since doing so would indicate more regret for the outcome of the attack. Third, the degree to which their satisfaction with the attack plan changes from before to after they learn that civilian casualties have been inflicted should be larger for those in the unmanned, high-selectivity treatment.

The top panel of table 6.2 describes the measurement scale and mean values for these variables across the four treatments.[18] Asterisks indicate if the mean values for the moderate and low-precision treatments are statistically distinguishable from the mean value for the unmanned, high-selectivity treatment. Respondents in this treatment group preferred higher levels of compensation. These differences in means are not large—the mean value for the high-precision treatment is $30,000, while that for the other two treatments falls between $25,000 and $30,000—but they are statistically significant. Recall as well that respondents were "anchored" at a value of $20,000. This anchoring likely pulled many respondents' choices in the direction of this value, as it was the modal choice for respondents in all

TABLE 6.2. Regret, Sympathy, and Civilian Casualties

Variable	Measurement	High Selectivity, Unmanned	Moderate Selectivity	Low Selectivity
Support for attack	1=Strongly approve, 4=strongly disapprove	2.36	3.01*	3.14*
Estimate of civilian casualties	1=Definitely, 6=Very probably not	3.63	2.40*	1.97*
Surprise at civilian casualties	1=Very surprised, 5=Not at all surprised	3.83	4.42*	4.53*
Compensation	1=No compensation, 9=$40,000 or more	7.06	6.54*	6.46*
Apology	1=Very important, 5=Not at all important	1.87	1.98	2.06
Satisfaction change	1=Very satisfied, 5=Very dissatisfied	–.63	.09*	.09*
		High Selectivity, Manned	**Moderate Selectivity**	**Low Selectivity**
Support for attack	1=Strongly approve, 4=strongly disapprove	2.32	3.01*	3.14*
Estimate of civilian casualties	1=Definitely, 6=Very probably not	3.56	2.40*	1.97*
Surprise at civilian casualties	1=Very surprised, 5=Not at all surprised	3.90	4.42*	4.53*
Compensation	1=No compensation, 9=$40,000 or more	6.32	6.54	6.46
Apology	1=Very important, 5=Not at all important	1.79	1.98†	2.06*
Satisfaction change	1=Very satisfied, 5=Very dissatisfied	–.59	.09*	.09*

* $p < .05$; †$p < .10$.

three treatments. The general support for this level of compensation, and the interest in paying more to victims of precise attacks, is itself an important finding. The United States has typically paid less to civilian victims of negligence—around $2,500—than the $20,000 anchor point that respondents gravitated toward. This suggests a general aversion to attacks on civilians and a feeling that they are a serious injustice against the victims.

There were no discernible differences between the treatments in the importance attached to issuing an apology. This is surprising; one might expect that those more willing to offer compensation would also be more willing to offer an apology. One possible explanation is that respondents viewed an apology as "cheap talk" and were willing to support an apology but not willing to support large payments to victims' families, which impose a monetary cost on the government and also might suggest a greater degree of culpability than verbal apology alone. It is also possible that respondents wanted to show sympathy for the victims of attacks, but without admitting to fault or misconduct by the American military. Payments to civilian victims are often given with the caveat that they are meant to express sympathy or condolence without being an admission of guilt.[19] The third measure, satisfaction change, is calculated by subtracting each respondent's level of satisfaction after learning the outcome of the attack from his or her response to the identical question asked before learning the outcome. One would anticipate that respondents primed to expect no civilian casualties would exhibit a negative value for this variable. This is the case, and these changes are significantly different from satisfaction change in the other two treatments.

Table 6.2 also reports mean differences for the remaining three variables, which serve as checks on how effective the information provided to respondents was in influencing their attitudes. Respondents assigned to the unmanned, high-selectivity treatment should be more willing to support the attack plan presented to them, since there is no risk to military personnel and little risk to civilians. These respondents should also provide lower estimates of the likelihood of civilian casualties and exhibit more surprise when informed that such casualties resulted from the attack. All three of these expectations are borne out in the data. This suggests that the treatments were effective in shaping the ways respondents thought about the value and consequences of the attacks. As important, it is experimental support for the supposition that the drone technology shapes individuals' assessments of its desirability and likely consequences.

The bottom panel of table 6.2 compares the high-selectivity, manned treatment with the moderate- and low-selectivity treatments. The mean values for the three manipulation checks are nearly identical to those for the unmanned selectivity treatment, suggesting that the experimental conditions produced similar reference points for respondents in both treatment groups. The fact that respondents in the high-selectivity, manned treatment experience statistically significant declines in their satisfaction with the attack plan after learning that it produced civilian casualties indicates that they also experience regret over civilian casualties. Compared to the moderate- and low-selectivity treatments, respondents in the high-selectivity, manned treatment were more willing to offer an apology but not willing to pay more compensation. This is the opposite of the pattern we see in the top panel of table 6.2 and is consistent with the logic of our second expectation. Even the remote possibility of military casualties reduces respondents' willingness to make costly payments to compensate for civilian casualties. Instead, they are now more willing to offer a less costly apology. This suggests that respondents in this condition continue to feel some regret for civilian casualties, but less than when the precision weapon is an unmanned platform.

Finally, we also compared the mean values of these variables for respondents assigned to the unmanned and manned high-selectivity treatments.[20] This allows us to assess how the presence of an onboard crew influences respondents' attitudes and preferences regarding high-selectivity attacks. This comparison is an important one because most of the aerial attacks carried out by the United States against militant targets meet our definition of selectivity, regardless of whether they are launched from drones or from strike aircraft. Respondents assigned to these two treatments did not differ systematically in their support for the attack plan, estimates of or surprise at civilian casualties, willingness to offer an apology, or surprise at the outcome. This is consistent with our findings across earlier chapters, as well as in other research, that the public perceives few differences between drone and air strikes.[21] However, those assigned to the unmanned, high-selectivity treatment were more willing to support compensation payments to the families of killed civilians than were those in the manned, high-selectivity group. Recall that this comparison holds the likelihood of civilian harm constant across the two treatments, suggesting that some mechanism other than selectivity is driving this difference. One possibility is that respondents were willing to offer more compensation for a drone strike because they believed that this platform was more likely to lead to civilian casualties. But the fact that estimates of such casualties across the two treatment groups did not differ significantly casts some doubt on this proposition. The key difference across these treatments is the presence of an onboard crew, which leads participants to support higher compensation payments.

Accidental Victims of War

The previous experiment indicates that disappointment regarding civilian harm increases as expectations about its likelihood declines. From one perspective, the finding that civilian casualties substantially reduce support for the use of force is surprising. In their detailed study of the issue, Larson and Savych argue that the public attaches a high priority to avoiding civilian harm when military force is employed, but the actual number of civilians killed or wounded by American military force seems to exert little influence on their attitudes. The reason is that many Americans have a high level of trust in the military, and believe that military leaders and planners routinely take considerable care to minimize the chance that civilians will be harmed when force is employed. This leads the public to view civilian casualties that result from the use of force as "unavoidable accidents of war."[22]

The regular use of drones opens up new opportunities to minimize the chance of such accidents. Drones' selectivity should reduce the chance of harm to civilians. For example, the Hellfire missiles fired from Predator and Reaper drones are laser-guided munitions that are far more accurate than "gravity" bombs. As important, drone operators are able to collect real-time intelligence on potential targets through video feeds and monitoring of cell phone, radio, and other communications, allowing attacks to be called off it they place civilians in danger. This sets drones apart from the manned aircraft and attack helicopters that can use the same munitions but that cannot carefully screen potential targets prior to launching an attack to confirm that they are enemy combatants.

Attempts to avoid causing civilian casualties with more conventional, less selective weapons often fail because they cannot be targeted with great accuracy or because the intelligence used to select targets is obsolete by the time an attack is launched. The argument that the public views civilian casualties as "unavoidable accidents" can be extended to cases where drones are available. The public should expect the military to take advantage of the selectivity of drones to launch attacks only when there is a high probability that civilian casualties will not occur. Rules and procedures about decisions to launch attacks should draw on the weapon's ability to integrate multiple sources of intelligence and to time attacks to minimize the possibility of killing noncombatants. When the abilities of such weapons to avoid civilian harm are not fully exploited, we can expect individuals to offer less support for the use of force.

Much of the controversy surrounding drone strikes centers on this issue. Until early 2008, the United States only launched drone strikes when they thought they had reliable intelligence on the identity and location of high-profile terrorists. After this date, procedures were changed so that the targets of drone strikes could include rank-and-file insurgents instead of leaders and so that strikes could be made against groups of armed men whose location and behavior indicated they were likely to be insurgents even if their identities were not known. While such "signature strikes" were only meant to be authorized if there were no civilians near the target area, many critics argued that this stricture would be difficult to implement. Many people in the area of Pakistan where most drone strikes occurred, the Federally Administered Tribal Areas (FATA), including those unconnected to insurgent groups, possess arms and travel in groups for greater security.[23] This meant that members of local militias, criminal organizations, and others who did not target the United States and should thus be considered noncombatants under international humanitarian law, likely were targeted by drones. Finally, defining and identifying "civilians" is often difficult in counterinsurgency campaigns. Insurgents deliberately conceal their affiliation with armed groups, by for example refusing to wear uniforms and by living and circulating among civilian populations.

According to their critics, signature strikes violate the norm of distinction.[24] Distinction is a key element of just war theory that governs the use of force during hostilities.[25] International humanitarian law, which is based in part on the ethical tradition of just war theory, governs the treatment of combatants and noncombatants during hostilities, and has been codified in a series of key treaties including the 1907 Hague Conventions, the Geneva Conventions of 1949, and two Additional Protocols to these conventions adopted in 1977.[26] The norm of distinction holds that combatants must always distinguish between military targets, such as enemy military personnel, equipment, and structures, which may be targeted with lethal force, and noncombatants, who cannot be deliberately targeted unless they are directly participating in hostilities. Importantly, international humanitarian law requires that those who "plan or decide upon an attack shall . . . do everything feasible to verify that the objectives to be attacked are neither civilians nor civilian objects" and, if there is "doubt" about the status of a potential target, "that person shall be considered to be a civilian" until the attacker can acquire "additional information and if need be give orders for further reconnaissance."

What we know about the practice of signature strikes from drones seems to violate these injunctions. Such strikes are carried out when there is uncertainty about the combatant status of the targets, while the failure to utilize drones' capacity to loiter for long periods and to collect real-time intelligence appears to contradict the need to "do everything feasible" to determine the status of the target as well as to seek additional information before attacking.[27] A reasonable extension of the logic of counterfactual thinking, then, is that individuals expect political and military authorities to exploit the opportunities that drones create for minimizing the risk to civilians when force is employed. Support for the use of force should decline when the opportunities are forgone.

This expectation provides an opportunity to assess two additional concerns about the experiment presented earlier in this chapter. First, while many individuals believe it is important to minimize potential harm to civilians when using force, there is considerable evidence that much of the public is not well-informed about domestic and international politics.[28] Some individuals may favor a policy of using force if it is implemented in a way consistent with just war theory and international humanitarian law, but are unfamiliar with the standards and principles to apply, such as that of distinction. In such cases, telling people that a policy is consistent with international legal principles may influence their preferences in particular situations.

Individuals who lack specific knowledge about a particular policy often adopt the positions of elites who share their underlying political preferences or independent third parties who are seen as providing unbiased information.[29] One study, for example, finds that members of the public are more likely to oppose policies when informed they violate international law.[30] Subsequent research has found that it is the *precision* with which international law defines an act as legal or illegal that matters, with the public more likely to oppose actions that international law clearly and precisely identifies as illegal.[31] If this is the case, providing individuals with credible information that an attack with drones violates international law should reduce support for the use of force compared to cases where such information is not provided.

Civilian protection is not the only concern of individuals, however. We might expect that some would be willing to accept a higher risk of civilian casualties in cases where the group targeted with attack poses a greater threat to the United States. This differs from civilian victimization, which is the deliberate and intentional infliction of harm on civilians as a strategy of conflict. Instead, as discussed earlier, individuals balance the benefits from the use of force against the costs. The key benefit, of course, is weakening the enemy; one of the costs is harm to civilians. It seems reasonable to expect, then, that attacks placing civilians at risk should receive more support if the target directly threatens the United States than if it does not.

Responses to Accidental Attacks: A Survey Experiment

We conducted a second survey experiment to assess this third expectation. As with the first experiment reported earlier in this chapter, respondents were over the age of eighteen, located in the United States, and recruited online. They answered questions about themselves and were then randomly assigned to read a hypothetical news story describing plans by the United States to launch drone strikes on suspected insurgent bases in Yemen. After reading the news stories, respondents were asked if they strongly approve, somewhat approve, somewhat disapprove, or strongly disapprove of the planned drone strikes. (The appendix to this chapter includes the texts of the treatments, the survey instrument, information about assignment and demographic characteristics of the sample, and robustness checks using ordered logistic regression models.)

The news stories were the treatments in this experiment, and they systematically varied three elements. In half of the treatments, labeled "high distinction," the attacks are described in terms consistent with the norm of distinction, stating that American authorities have the capacity to closely monitor the insurgent bases and are "confident" they can time their attacks to avoid civilian casualties. In the other half of the treatments, labeled "low distinction," the vignettes state that despite attempts to monitor the insurgent bases, the attackers cannot be sure that all victims will be combatants.

The second factor is threat, operationalized as the degree to which the insurgents directly threaten the United States. Half of the treatments ("high threat") describe the insurgents as having targeted the United States, while the other half ("low threat") state that the insurgents plan to attack only targets in Yemen, an ally of the United States. The third factor concerns international law. One-third of the treatments make no mention of the legality of the planned attacks. One-third state that the attacks "would violate international law." The remaining vignettes state that the attacks "might" violate international law and that independent experts are divided on this question. This captures the precision dimension of international law, with the last group of treatments offering less precise information regarding the legality of the planned targeted killings.

After reading the news stories, respondents in the experiment were asked if they strongly opposed, somewhat opposed, somewhat supported, or strongly supported the attack plan. The least supportive response was assigned a score of 1 and the most supportive a score of 4. Figure 6.1 summarizes responses in each of the twelve treatments. The bottom panel treatments vary the degree to which the attacks are consistent with the norm of distinction and the threat the target poses but make no mention of their legality. Comparing responses across these four treatments allows an assessment of the specific expectations that support will decline when the attack is less consistent with the norm of distinction and the threat to the United States is greater.

The data is consistent with these expectations. The values of this variable are lower for the low-distinction treatment than the high-distinction treatment, and these differences are statistically significant. The same is the case for threat; support for the use of force is lower when the insurgent group targeted with drone strikes has not itself launched attacks against the United States. Interestingly, the effect of distinction is considerably larger than that of threat, and the difference between the low-threat, high-distinction and the high-threat, low-distinction treatments is statistically significant. Furthermore, these factors have an additive relationship to the dependent variable. Combining a low threat with low distinction leads to a larger decline in support than do either of these factors alone. The differences between the low-threat, low-distinction and both the low-threat, high-distinction and the high-threat, low-distinction treatments are statistically significant.

The remaining treatments in figure 6.1 assess the effect of international

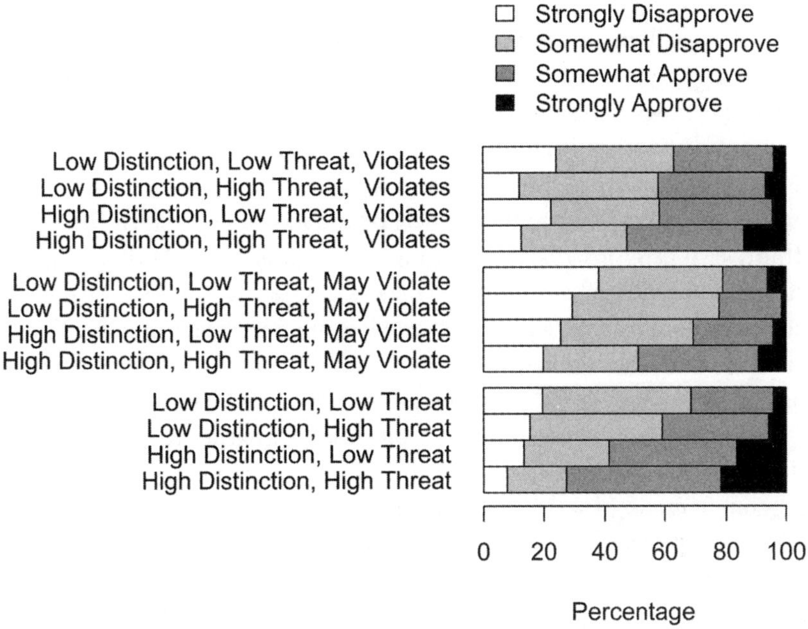

Fig. 6.1. Distinction, threat, and law

legal commitments on support for drones. The treatments in the top panel of figure 6.1 are identical to those in the bottom panel discussed above, and also state that the drone strikes violate international law. Mean values for these treatments are all lower than for the corresponding treatments that make no mention of international law, and these differences are statistically significant. This indicates that respondents are less likely to support attacks that clearly violate international law. Note that the effect of violating the norm of distinction is independent of the contribution of international law. In other words, violating the norm of distinction alone reduces support for the use of force. Informing respondents that the attack violates international law also reduces support, but these effects are not contingent on each other. The final four treatments state that the attack *may* violate international law. There is a much less consistent pattern here. The possibility of violating the law leads to reductions in support in only two of the four comparisons. This suggests that the precision of international law is particularly important; respondents are more strongly persuaded when the legal violation is more precise.

Looking across all of these differences, three patterns emerge. First, respondents' willingness to support an attack declines when it is less consistent with the principle of distinction. Second, support also declines when the target poses a less direct threat to the United States. Importantly, the effect of distinction is larger than that of threat, indicating that respondents are, on average, unwilling to abandon their concern with civilian protection when doing so might increase the security of the United States. Third, legal precision matters; treatments that clearly specify that an attack violates international law have lower levels of support than those that make no mention of international law, while less precision regarding the legality of the planned drone strikes do not produce consistent differences from similar treatments that do not mention international law.

Some might be encouraged by the finding that attacks described as clearly violating international law consistently have less support. It is important to note, however, that there is considerable controversy among real-world experts over the legality of targeted killings outside of war zones. Actual news stories that address this issue are more likely to describe attacks as potentially inconsistent with international law and will thus more closely resemble the vignettes used here. This suggests that the practical effect of international law on attitudes may be small. Instead, attitudes are more predictably influenced by consistency with the underlying norm of distinction.

Conclusion

Drones are better-equipped to precisely hit their targets than other aircraft, and they reduce the risk of harm to military personnel. These characteristics lead some to worry that the development of these weapons will decrease the costs and risks involved in using force. If this is the case, it should operate with particular effect in democracies such as the United States. Elected leaders must worry that their decisions to use force can backfire against them, as unsuccessful missions or high levels of military casualties reduce their domestic political support. The fact that President Carter was not re-elected in 1980, for example, has been blamed in part on the failure of the mission he authorized to rescue American hostages being held in Iran.[32] Presidents Johnson and George W. Bush saw their popularity decline as military casualties mounted in the Vietnam and second Iraq wars, respectively. The introduction of drones on a large scale, then, might make it politically easier for presidents to authorize the use of force, safe in the knowledge that drones are more likely to achieve their goals and less likely to lead to the death or capture of service members. If drones make war politically cheaper and easier, we might expect leaders to resort to force more often and more recklessly.

At the same time, though, drones also increase the capacity to avoid harming civilians. The ability to collect real-time intelligence and integrate it with other sources of information allows drones to accurately guide munitions to a well-defined target. This allows these platforms to be armed with smaller weapons that are less likely to create collateral damage. We have suggested that this development creates novel political constraints on the use of force. Individuals' expectations adjust with technology; as weapons become more capable of avoiding harm to civilians, people expect their political and military leaders to utilize this capability and adjust their expectations regarding civilian casualties.

The first experiment in this chapter demonstrated that respondents primed to believe that weapons used in an attack are accurate express more disappointment and regret over civilian casualties than do those primed with information about less accurate weapons. This difference exists even though the outcome of the attacks is described in identical terms in both treatments; in other words, respondents in the experiment with high expectations for avoiding civilian harm are considerably more disappointed when this outcome fails to materialize. While drones increase support for the use of force by increasing the chance of a successful mission and reducing the likelihood of military casualties, they also generate heightened concern about and sensitivity to civilian casualties. The development of drones, then, does not have a straightforward or linear effect on support for the use of force.

The second experiment extends this insight. It compares attack plans that do and do not fully exploit drones' capacity for minimizing civilian casualties. The former treatment receives more support from respondents in the experiment, indicating that they expect planners to take available steps to avoid civilian harm. This difference increases when respondents are told that the attack would violate international law. This suggests that human rights campaigners and others who view drone strikes as violating international norms for the protection of civilians would be well-served to emphasize this point, as doing so may shift the terms of the debate about development and use of such weapons. Furthermore, the difference in levels of support between attacks that do and do not fully distinguish civilians from combatants remain even when the threat to the United States is described as more direct. This suggests that the concern for civilians is not superficial, as respondents in the experiment remained opposed to less discriminate attacks in such treatments.

Finally, the first experiment found important differences between attacks from manned aircraft and from unmanned aircraft. Concern about harm to civilians declined when the attack involved an aircraft with a pilot, even though the insurgents were described as unable to attack this aircraft and the outcome in terms of civilian harm was identical to the unmanned treatment. This indicates an important limitation on the findings of this chapter: worry about killing civilians appears to decline as the risk to American military personnel increases even modestly. One implication of this is that greater use of unmanned precision weapons in the future may heighten the political sensitivity of collateral damage. However, it also suggests that this effect could weaken if opponents of the United States themselves develop more selective weapons. If this provided them with the ability to more effectively target American military forces with harm, or to launch attacks on the territory of the United States, concern about the civilian harm inflicted by American precision weapons might dissipate.

CHAPTER 7

Conclusion

Democratic leaders are not free to make war whenever and however they choose. They respond in part to citizens' wishes, or they risk the consequences of a public backlash when enacting new policies or running for election. American citizens have played an important role in constraining executive power by protesting wars, refusing to fight them, and attempting to remove leaders from office.[1] Citizens are equally important when giving a mandate for their leaders to wage wars without fear of electoral punishment, or even rewarding violence with higher approval ratings. Existing research on public opinion during war helps to identify the causes for antiwar and prowar sentiments, but advancements in military technology and possible shifts in citizens' values over time make it vital to continually reassess the influences on support for war. The twenty-first century has thus far been marked by a seemingly unending string of American conflicts, from major operations in Afghanistan and Iraq to smaller interventions in Libya, Yemen, Somalia, Syria, and Pakistan. These raise legitimate concerns that changes in the conduct of war or in civic engagement with military affairs could fundamentally alter the conditions under which politicians may initiate wars.

Drones stand at the forefront of the debate over whether wars are becoming dangerously risk-free for the United States. They are the latest phase of a revolution in military affairs that put greater emphasis on air power, increased selectivity, and on preventing harm to American soldiers.[2] Drones take the United States' search for "useable weapons"[3] that allow it to project power at relatively low cost to its most extreme conclusion by making it possible to launch strikes while keeping pilots so far away from the battlefield that they are invulnerable from enemy fire. For the first time in history, a machine can act as a reliable proxy for a human soldier or manned vehicle, rather than as a single-use weapon that detonates on impact. This raises legitimate fears of wars becoming so easy that the public—and maybe even soldiers themselves—will lose any sense of the costs of violence and become less restrained.

Drones' selectivity is equally revolutionary. Although the United States has pursued precision warfare for decades and uses other weapons, such as laser- and GPS-guided munitions, to achieve accurate strikes,[4] drones mark another step forward by making it possible to launch much quicker attacks using real-time surveillance. Greater selectivity provides the technical capacity for waging wars more effectively, but it also risks incentivizing war by making drone strikes seem like the ideal tool for killing opponents even when this is not essential to national security.

Thus far, the debate over whether drones create unique dangers that would make them morally questionable, or whether they are just another type of military hardware that will fit into established patterns of fighting, has been primarily conducted in moral and legal terms. There is little empirical research to substantiate the claims offered by either side, especially when it comes to their predictions about how public opinion will change.[5] To some extent, this framing is appropriate. It is impossible to test claims about issues like whether the extreme asymmetry of drone warfare is inherently immoral or whether drone strikes contravene martial virtues that militaries depend on for normative orientation. These types of issues are best approached with moral and legal analysis. However, many of the claims being made by those on both sides of the debate rely on empirical assumptions, and throughout the book we have sought to advance this debate by describing the most compelling versions of these arguments and testing them with survey experiments.

Perhaps the most common worry coming from critics is that drones will lower inhibitions against fighting and lead to a proliferation of violence. Those who support the use of drones contend that this fear is overstated and that drones are employed in much the same way as manned aircraft or cruise missiles. Our results show that both sides are partially right and partially wrong: the availability of drones does sway public opinion, but not as much as many critics have said and with significant influence from other variables. Everything else being equal, the American public is more willing to fight a war with drones than with soldiers who are physically present on the battlefield. Attacks involving manned aircraft were also more popular than ground forces but less popular than drones.

This pattern of ranking the three options holds several important insights. First, it shows that the American public is sensitive to military casualties and that this influences attitudes toward military action. Politicians hoping to mobilize support for fighting could therefore benefit from not committing ground forces and employing drones instead. Second, because the pattern is consistent across missions designed to achieve different principal policy objectives, our results suggest that the preference for drones will persist even as the strategic context and reasons for fighting change.

Determining whether the increased support for war substantiates claims that drones make it easier to fight raises the issue of their substantive effects size that we have addressed throughout the book. Drones can be said to have a large or small substantive influence on support for war depending on one's prior beliefs and what background literature is used to contextualize the findings. The effect of using drones is about as large as other factors that are known to influence attitudes toward war, such as the prospect of success or the likelihood of civilian casualties. These comparable substantive effects demonstrate the importance of including drones in public opinion research that gauges attitudes toward war.

The substantive consequences of drone technology look much smaller when interpreted in light of fears that drones will dramatically lower the threshold for fighting, cause a rapid increase in American militancy, or undermine democratic accountability. No single variable that we tested could be said to have such a strong effect on support for war—even drones. It is more accurate to say that drones can play a part in building the overall case for war, but that they lack the power to independently cause major shifts in American foreign policy or in the public's ability to hold leaders responsible for how military force is used. For certain individuals, drones clearly lower inhibitions against using force, and it is possible to imagine a situation in which choosing drones over competing methods could tip the balance of public opinion when the electorate is sharply divided, or a situation in which drones could form part of a broader effort to build public support for fighting that includes manipulating other variables (such as framing a conflict in terms of counterterrorism or showing that there is a high likelihood of success). Those who defend drones would be wrong to argue that having these weapons platforms available makes no difference to the United States' propensity to fight, yet they are correct in saying that the degree of increased support is probably too low for drones to cause a major shift in the incidence of wars.

Public opinion does not directly influence policy, and leaders may to some extent ignore public opinion when declaring wars if they think they will be vindicated in the long run.[6] The primary threat from low approval ratings is losing the next election,[7] and this is generally a more distant concern that can in principle be countered by quickly achieving a good outcome, shifting the electorate's attention onto other issues, or evading blame for the decision.[8] Even at the lowest moments during the occupation of Iraq, President Bush was steadfast in asserting that his decision to invade the country would be vindicated in the long term by a successful outcome.[9] It is easy to imagine other leaders embracing short-term risks in the same hope of being ultimately proven right. When applying our findings in practice, it is essential to be aware of the relative autonomy politicians have to make unpopular decisions and the electorate's inability to directly control use of force decisions. This explains why states that have drones continue to rely on other means of attack in some instances, even when these are apt to be less popular. Politicians may simply decide that it is strategically and politically preferable to risk some small losses of support when prioritizing other goals.

It is also important to look at partisan identities when interpreting our results. We found in chapter 2 that while Republicans were more willing to endorse the use of force in general, members of both parties preferred attacks by drones to the use of ground troops. This contrasts with other findings about the polarization of support for American military interventions over the past two decades.[10] Support for drone strikes over ground troops exhibits fewer partisan differences. As we suggested in chapter 2, this may explain why President Obama, elected in 2008 on a platform that emphasized winding down the large-scale American troop presence in Iraq and Afghanistan, continued and even expanded his predecessor's use of drone strikes against militants in Pakistan and elsewhere.

Our experiment involving principal policy objectives (PPOs) showed that members of the American public make prudential calculations when it comes to judging when and how to launch military interventions. Variations in the PPO led to significant divergences in enthusiasm for fighting regardless of the type of attack that was chosen. Counterterrorism was seen as being far more important than the other three PPOs that we tested, and it received more support than alternative PPOs regardless of the attack method. Foreign policy restraint came next, followed by internal political change—a sign that respondents were less concerned about threats from conventional militaries and more inclined to stay out of conflicts that do not seem to pose an immediate threat to US national security. Humanitarian interventions received the least support. When this is compared against previous research on PPOs,[11] it seems that the preferences for certain PPOs over others changes over time based on the perceived salience of the PPO for the current national security context. Terrorism was clearly seen as the biggest threat, as it was easier to build a case for fighting terrorists than other enemies. As security priorities change, the relative attractiveness of the PPOs may do the same.

For some PPOs, drones were consistently more popular than attacks involving ground forces and air strikes, while air strikes were more popular than ground attacks. This produced a step pattern for each means of attack across the PPOs that suggests a consistent ranking of preferences, which further reinforces the sense of a public making reasoned calculations based on the costs and benefits of military action in each scenario. Thus, our findings not only support the "prudential public" thesis by replicating some previous findings of work on PPOs but also identify another dimension of the public' strategic calculus. This demonstrates the importance of recognizing the public's competence and ability to make decisions when it comes to foreign policy issues, and it indicates that any theory of public opinion will need to start by recognizing that the public is not always easily swayed by elites or who have no sense of what is at stake in America's wars.

It is difficult to predict how successful drones are in actually achieving the objectives set out in the PPOs. Proponents of drone use contend that their ability to kill senior figures in terrorist organizations and militant groups with selective attacks will make them more effective than alternatives, while their critics argue that drones create resentment that will intensify hostilities and encourage enemies to deviate from the norms of war—especially by terrorizing civilians with constant surveillance. At present, we lack clear evidence to show which side is right, especially when drones have only been employed in the kinds of counterterrorism operations that are notoriously difficult to win regardless of how they are conducted. However, it is possible to test whether the public thinks that the chances of winning depend on how an attack is carried out.

Looking at the anticipated success of drone strikes compared to other attack types is vital for situating these weapons platforms in the context of existing public opinion literature. Gelpi, Feaver, and Reifler have done some of the most influential and rigorous research on public support for war and find that anticipated success plays the decisive role in determining the overall level of support.[12] If drones caused a major change in the predicted outcome of a conflict, then they could likewise cause a major shift in levels of support for fighting. We found that drones were not seen as being any more effective than air strikes or ground forces. The attack types were almost interchangeable when it came to predicted success, which means that drones will not have a strong influence on this variable that was so important in previous efforts to account for Americans' evaluations of wars.

The concern over drones making war easier and thereby tempting the public into supporting more wars or treating war as a quick fix is understandable. After all, moral hazard is a well-documented phenomenon, and the underlying reasoning fits with the finding that choosing drones over other means of attack leads to an increased willingness to fight. This concept is also important for evaluating drones through the lens of just war theory. The principle of last resort requires that belligerents pursue all available means of peaceful conflict resolution before they fight. One reason why moral hazard is threatening is that it would be immoral based on just war theory for a belligerent to rush into a war when promising alternatives exist. Our experiment testing the effects of moral hazard looked at whether drones would be more popular than other means of attack when military ventures were risky and nonviolent options were available. We found that in this context the effects of moral hazard did not appear and that respondents consistently wanted to take a nonviolent approach before resorting to drones or other weapons. This further indicates that support for drones, or for any types of military force for that matter, coheres with the principle of last resort and that public opinion will urge politicians to explore nonviolent means of conflict resolution when these are available.

Our experiments involving expectations of military casualties showed the extent to which pilot invulnerability alters attitudes toward prospective attacks, but these do not directly address the effects of increased precision. To evaluate this other defining characteristic of drones we looked more closely at how respondents felt about the civilian casualties that are supposed to be reduced as weapons become more precise. The experiment involving counterfactual thinking showed that predictions about weapons being precise or imprecise prime responses to the attack outcomes. Those who read stories promising low numbers of civilian casualties reacted more negatively to the results than those who were primed to expect high casualties, even though both groups read the same story about the results of the attack. Here we see that the American public is sensitive to foreign civilian losses and that overall judgments about the permissibility of attacks will be framed with these in mind.

The experiment comparing high- and low-distinction attacks provided additional support for this finding. In that case we found that respondents were less supportive of drone strikes when there was a higher likelihood of harming innocent people and that this had a greater influence on their judgments than the level of threat posed by the target. Even more importantly, that experiment showed that concern over civilian welfare was higher when there was no risk to American soldiers. One upside of using drones and eliminating the risk to soldiers is therefore that the public will give more attention to civilian suffering and formulate judgments about attacks with that in mind.

Policy Implications

Even though the increased support for military operations involving drones compared to those involving ground forces likely has a modest influence on building enthusiasm for fighting, the prospect of drones increasing aggression is a potential danger that should be taken seriously. One response to this would be to impose stricter regulations that can restrain aggression. Buchanan and Keohane propose an international drone regulatory regime to "enhance accountability through transparency and publicity and, in countries where this is feasible, to mobilize domestic constituencies to support compliance with the regime."[13] According to this proposal, states that use drones would create their own supervisory agencies. Each state would likewise monitor the others to ensure that drones were used in accordance with the established provisions, and in particular that prospective strikes were discussed and agreed on before being launched. In the event that attacks have to be carried out quickly, Buchanan and Keohane say that states would need to follow procedures for gaining *post facto* authorization. This concession to practical necessities would allow states to protect vital national security interests while still promoting transparency by forcing states to explain the rationale for attacks and publicize the outcomes.

Other commentators likewise advocate more regulation and oversight of drone strikes, though usually without going into as much detail about how these would work. For example, Boyle argues that the United Nations should form an investigatory body to monitor drone use and sales.[14] This would lack the power of the regulatory regime Buchanan and Keohane advance, but might be a more realistic proposal and would certainly help to generate greater attention to drones from the international community. At the very least, such a body could call attention to when drones are being misused and attempt to mobilize international public opinion against the offending belligerent.

One point of agreement between the various regulatory proposals is that the United States needs to take a leading role in implementation. They contend that it must act quickly while it is in a position of strength—because it will soon lose its monopoly over drones and the leverage that goes along with that special status. In other words, they argue that it is in the United States' long-term interest to introduce some regulatory framework for drones and that failure to do this in the near future could reduce the chances of ever developing effective regulations. The question now is whether the United States would actually support regulating drone strikes at this time. Our experiments did not inquire into respondents' feelings about regulation, yet our results do shed some light on Americans' attitudes toward drones. The greater popularity of drones compared to ground forces and aircraft indicates that Americans see these weapons platforms as being attractive, which casts doubt on whether they would want to relinquish the ability to employ drones unilaterally and without seeking approval. Respondents' low confidence in international organizations, aside from NATO, is further reason to doubt that the American public would support a regulatory regime grounded in some international body.

Whether an increase in the public's willingness to assent to wars is a good or a bad outcome is a matter of debate. For pacifists, anything that facilitates war, even slightly, is dangerous and objectionable. Many commentators who criticize drones likewise imply that any increased freedom to initiate wars is a bad outcome. As we pointed out in chapter 3, Beauchamp and Savulescu raise the possibility that drones could make it easier to initiate benevolent humanitarian interventions in which the intervener may be unwilling to provide assistance unless it can be done cheaply.[15] From this perspective, drones could make atrocity prevention more politically palatable to the American public. Along the same lines, many commentators who decry the public's casualty aversion or America's ability to credibly deter rivals when so many citizens oppose fighting may welcome drones. From their perspective, drones may have strategic benefits, such as their ability to maintain a balance of power or to punish rogue states, that might help to maintain security by preventing the outbreak of large wars.

New regulations may also be adopted unilaterally by the United States with the goal of promoting accountability to the American public. One of the most urgent concerns is that drones have been used to kill American citizens without trial. The killing of Anwar al Awlaki marked a critical turning point in the history of drone warfare. He was not the first American killed by a drone—that distinction belongs to Kemal Darwish, a suspected al Qaeda recruiter who was inadvertently killed in Yemen in 2002—but he was the first American citizen to be *targeted*. The attack against al Awlaki, which came on September 30, 2011, while he was living in Yemen, marked an important shift in the War on Terror.[16] Selecting him for attack set a precedent of treating American citizens involved in terrorism as legitimate targets for extrajudicial killings.

Samir Khan, the editor of al Qaeda's *Inspire* magazine and an American citizen, was killed alongside al Awlaki, though he was reportedly not the target. Al Awlaki's son Abdulrahman was killed two weeks later during a strike that Obama administration officials said was directed against Ibrahim ad-Banna, a senior al Qaeda commander from Egypt. Four more Americans were killed over the following years: Jude Kenan Mohammad, Ahmed Farouq, and Adam Gadahn each had links to al Qaeda. Warren Weinstein, a 73-year-old US Agency for International Development contractor, was killed inadvertently when he was being held hostage.

The Justice Department ruled that targeting American citizens like al Awlaki without a trial is legitimate because their involvement in terrorist plots constitutes a threat to national security. However, critics have argued that extrajudicial attacks violate Americans' right to due process, especially when they take place in areas where the United States was not actively waging wars, as was the case in Yemen.[17] An urgent concern for bringing greater normative clarity to the strikes against Americans is establishing clearer guidelines about when the norms relating to war pertain and when the much different norms relating to law enforcement pertain. Throughout the War on Terror the US government has shifted opportunistically between employing war and law enforcement paradigms, causing confusion about how to evaluate drone strikes.[18]

Another important issue when it comes to promoting the responsible use of drones is ensuring that citizens and legislators have the information they need to make prudential calculations. Unfortunately, this information is often unavailable, especially with covert strikes carried out by the Central Intelligence Agency and Joint Special Operations Command (JSOC). Reports on the drone program suggest that these organizations are much less open about when and how their drones are employed than is the US Air Force.[19] The Obama administration took some steps toward increasing transparency by transferring control of drone operations away from the Central Intelligence Agency to the military, yet this has led to more operations being carried out by JSOC without any discernible improvements in terms of public oversight.[20] For citizens to make the kinds of use of force decisions we have discussed throughout the book, it is vital for the government to provide citizens with more information about prospective attack plans as well as to give more accurate reports about the consequences of attacks so these can be weighed in future decisions. Increasing the information available about drone strikes may be the easiest route to restricting the use of drones, as this would make it possible to activate public opinion as a more effective constraint.

Our experiments not only hold lessons for policymakers who are attempting to manage public opinion as they employ military force but also for those opponents of drone warfare that hope to build a consensus against them. Arguments against drones that are rooted in the logic of moral hazard would be difficult to substantiate empirically. Arguing that drones might increase support for fighting might also not be an effective argument because of the public's preference for not placing American military personnel at risk. Anti-drone activists' efforts would be best served by emphasizing the extent to which promises of low civilian casualties are not borne out in practice and by challenging the lack of transparency surrounding these casualties. These approaches provide a better opportunity for provoking outrage.

One of the core lessons for proponents and opponents of drones alike is the importance of transparency about drones and how they are used. In each experiment we imagine the choice to use drones as a fairly straightforward selection between several competing options in which respondents had a clear sense of what the objectives were and even some clues about the possible outcome of attacking. Citizens are never given such a clear choice about how and when to fight, but the experiments do mirror the way politicians and journalists discuss various attack options when debating whether to launch an attack. The ongoing discussion about how to respond to the war in Syria provides ample evidence of this. Proponents of intervention have unveiled dozens of different plans for resolving the conflict with a minimal loss of American lives, and these options have gravitated toward deploying a small contingent of American forces, attacking with manned aircraft, attacking with drones, or relying on a combination of these attack types.

Our experiments show that the public is prudent in its judgment about when wars are justified and what means of attack to employ when they have a fairly clear menu of options, yet our experiments assume that the public has all the relevant information to make informed decisions about initiating violence. Unfortunately, we cannot assume that this level of openness will exist in practice. The United States government has been reticent about revealing the number of drone strikes it launches or giving casualty estimates.[21] The lack of transparency interferes with the public's ability to monitor the effects of drone warfare and to make informed decisions between the policy choices available.

Clearer civilian casualty estimates would impose constraints on how politicians actually use drones by offering evidence that can inform judgments of whether promises of low civilian casualty rates are being met. Most independent estimates of the civilian casualties resulting from American drone strikes put these figures at less than a thousand throughout the Obama and Bush administrations. Any amount of civilian suffering is tragic and regrettable, but these numbers are much lower than those associated with major ground combat operations, as evidenced by the casualty figures from Afghanistan and Iraq. Drones may even inflict fewer civilian casualties per attack than special operations forces and cruise missiles.[22] But senior figures in the Obama administration, including the president himself, as well as other high-ranking figures in Congress, the military, and intelligence services, made lofty promises about drones only killing a few civilians.[23] Our results suggest that most Americans would be inclined to support drone strikes against terrorists, but that hiding the civilian costs of attacks is setting them up for much greater disappointment than if these costs had been acknowledged at the outset. That is to say, misleading promises of perfect precision could be counterproductive simply because these are misleading. Our counterfactual experiment demonstrates that more honest and realistic predictions about attack outcomes will be a more enduring strategy for maintaining public support.

Another implication when it comes to changing how the United States conducts wars is that respondents supported giving relatively large amounts of financial compensation to civilian victims. This is evidence that there is not only concern for civilian welfare but also a fairly widespread desire to repair harms inflicted on the innocent. These amounts were influenced by anchoring the value at $20,000, rather than at the $2,500 limit on payment amounts that was imposed throughout the wars in Iraq and Afghanistan, yet respondents were willing to go even higher than this and therefore showed that they genuinely wanted to help victims of misdirected violence. This suggests that the American public would be receptive to the higher levels of compensation that some critics of the existing compensatory payments have called for.[24]

Looking beyond payment specifically, the conjoint experiment in chapter 2 and the experiments we discussed in chapter 6 show that the American public has a complicated relationship with international organizations and the norms that are supposed to inform the conduct of wars. The conjoint experiment in chapter 2 found that Americans were less likely to support attacks when they were opposed by NATO and that they were also less likely to support attacks when the United Nations Security Council or human rights nongovernmental organizations expressed an opinion toward them, regardless of whether the opinion was positive or negative. This level of mistrust for international organizations, except for NATO, seems to indicates that Americans wish to act unilaterally. However, the experiments in chapter 6 showed that Americans respect the principle of distinction and that they were concerned about whether the attacks they authorized were in accordance with international law. From this we can surmise that Americans respect international norms even though they are suspicious of the organizations that are charged with upholding them.

The Future of Drone Innovation

Throughout the book we followed the common practice of using the term "drone" to refer to unmanned aerial vehicles like the Predator and Reaper, but really our experiments have focused on one of the many different types of weapons platforms that may be described as drones. UAVs are currently at the forefront of military innovation, but as we pointed out in the first chapter, armed forces around the world are creating dozens of drone variants, including machines that can operate on land and at sea. Future research will need to take up the challenge of gauging public opinion related to other types of drones, especially machines that operate in different terrains or with less human control. It is impossible to reach definitive conclusions about these until the machines are capable of being deployed in combat, at which point they will enter into the calculations relating to the use of force. However, it is possible to extrapolate some predictions based on our study of public support for current-generation UAVs.

We expect future UAVs to fit into roughly the same patterns we have described throughout the book. The close parallels between support for piloted aircraft and UAVs across most experiments indicates that public support for their use only varies slightly, despite major technical differences. Future UAVs are likely to be faster and better protected from enemy fire and perhaps more capable of operating in conventional combat roles. But if such a radical step as removing pilots from the aircraft only produces a slight effect on public opinion, it seems unlikely that improvements to drone design will have much impact. Our results indicate that any public opinion advantage drones may secure comes from pilot invulnerability and greater precision—characteristics that help to define drones as a class of weapons platforms and that will therefore persist across the many different models that may be introduced. So long as pilot invulnerability and selectivity remain the guiding themes of drone innovation, public opinion should continue along the patterns we have identified.

Most of the naval drones in development are designed for fleet security and other roles that would have them more removed from daily life for most civilians. Except for people employed in maritime industries, naval drones would be just as distant from public view as are UAVs. Those machines could also maintain a high degree of selectivity, or at least perceived selectivity, by operating away from populated areas where civilian casualties are likely. For these reasons, we would expect limited public opposition to naval drones and also little risk they will alter the public's calculations when it comes to initiating conflict.

We expect more variation in support when drones that would operate on land are introduced. These machines could appear to be more intimidating than UAVs because of their proximity to civilians. Daily contact with ground-based drones would certainly have a profound effect on people in contested areas where the machines would be used. However, their influence on public opinion within the United States will depend on how the machines look from a distance. Those with animal and humanoid appearances may be anthropomorphized, which could produce a sense of loss when they are damaged or destroyed, but this is unlikely to match sensitivity to human losses. For this reason, ground-based drones will maintain the theme of pilot invulnerability and offer the same benefits as UAVs when it comes to protecting American soldiers. The larger question will be whether these machines will be perceived as instruments of precision warfare. This will depend on how the machines are armed, how they are framed by the media and government officials, when they are deployed, and, most importantly of all, the track record they establish in combat.

The pattern of increasing support for drones over ground forces shows the extent to which reducing the risks to military personnel can elevate approval for wars. Other means of attack that we have not tested probably fall along this same continuum, depending on the extent to which they put Americans at risk. Deploying regular infantry units is probably less popular than the elite ground forces we discussed in our experiments, while cruise missiles are apt to have approval ratings that are much closer to those of drones. Although each attack type may have a slightly different position along the continuum, we expect that drones show the furthest extent to which approval for war can be increased by removing military personnel from danger. Because drones guarantee complete pilot invulnerability, it is impossible for any weapons platform to make American soldiers safer and thus reap additional public opinion benefits of preventing harm to military personnel.

Each of our experiments tests the use of drones in military contexts. Even the counterterrorism and humanitarian PPOs we discussed in chapter 4, which sit at the border between warfare and policing, were described in terms of an armed military intervention in a foreign country. One important question for future research will be determining what influences support for drones in domestic contexts. Since 2004 the United States Customs and Border Patrol has employed unarmed Predator B drones. These aircraft have been lent to law enforcement agencies within the United States, and many are attempting to acquire their own drones.[25] There has even been some speculation about arming drones with nonlethal weapons, though it seems unlikely that this possibility will be realized for some time.[26]

Drones in domestic law enforcement raise many unique concerns, such as whether these are consistent with privacy rights and whether they should be armed. It is vital to gauge public opinion regarding these drones as they are introduced, especially when so many are dual-use machines that can be employed in international and domestic conflicts. Such drones highlight a contrast between war and law enforcement and a confusion of the respective normative and legal paradigms that has also been characteristic of the covert targeted killing missions drones have been involved in throughout the War on Terror.[27] It will be important to see whether the public uses different criteria to evaluate drones in domestic settings, how comfortable they are with dual-use technologies, and whether drones could lead to more aggressive law enforcement tactics that mirror how drones have been used internationally.

Autonomous drones that are capable of selecting and engaging targets without direct human involvement present an especially important research challenge and are apt to generate a stronger reaction from the public than remotely piloted machines operating in new terrains or in different security contexts. Autonomous drones provide the same pilot invulnerability and selectivity that is characteristic of UAVs, but they also allow artificial intelligence systems to make, or at least contribute to, the decision of whether to kill. Much of the debate over the implications of drone warfare is directed at autonomous drones specifically,[28] and here there is a concerted effort to shape public opinion before these machines can be developed. Prominent scientists and inventors including Stephen Hawking, Steve Wozniak, and Elon Musk have spoken out against autonomous drones, while organizations like the Future of Life Institute, the Campaign to Stop Killer Robots, and the International Committee for Robot Arms Control campaign raise public awareness about autonomous drones and urge politicians to ban them. Popular culture also supplies an endless stream of movies, videogames, and television shows about autonomous robots threatening their creators and even precipitating the collapse of human civilization.

Removing or weakening human control over the use of lethal force would mark a radical change in how drones operate and would require careful analysis of resulting shifts in public opinion. We avoided taking on autonomous drones in this book because these are largely speculative at the moment, and the general public lacks any experience with what effects they would have in practice. Some work has been done to see whether members of the public are able to determine which human actors are responsible for the use of autonomous drones,[29] but more will have to be done to see whether these machines disrupt the influences on public opinion we have described. Given the substantial opposition to autonomous drones already evident and countless popular narratives involving killer machines, we predict that decisions to create or deploy autonomous drones would generate considerable public outcry. Such machines would almost certainly be less popular than remotely operated drones and would therefore undercut any public opinion benefits that UAVs might yield. This would make reliance on these machines a risky venture for democratic leaders.

Drone Proliferation

Drones are spreading far beyond the United States. It is difficult to say precisely how many countries are developing UAVs, as such information is often carefully guarded. However, estimates suggest that as of 2012 at least 11 countries had developed UAVs that were capable of lethal attacks.[30] Aside from the United States, Israel and the United Kingdom are the only countries that have made extensive use of UAVs in combat. Israel used aerial drones at least as early as 1982, when they were sent over the Bekaa Valley as decoys during the Lebanese Civil War.[31] Operations in the urban environments of Gaza and the West Bank prompted Israel to use drone surveillance as a means of acquiring targets that could be attacked by manned aircraft or ground forces. As in the United States, reconnaissance gave way to offensive missions, and drones were soon being used to locate and attack suspected terrorists. Between 2000 and 2014 Israel conducted around 270 targeted killings in the Palestinian territories, killing around 455 people.[32] Its IAI Heron and IAI Eitan UAVs are similar to the American Predators and Reapers and are used for the same kinds of missions to monitor and kill suspected terrorists, as well as for reconnaissance flights over neighboring states.[33]

British involvement with drones developed through joint operations with the United States. These began in Iraq, with the first reported attack by an RAF pilot taking place in 2004. In some instances American and British pilots have collaborated to attack specific targets. For example, two British citizens working with Islamic State in Syria were killed by a combined team of American- and British-controlled Reaper drones in 2015.[34] The United States is reluctant to give armed drones to its other allies, but it has provided unarmed Predator drones and helped to train pilots for other

NATO member states, thereby giving them some capacity for operating these machines independently. And soon foreign support may be unnecessary, as members of the European Union have begun collaborative work to produce their own UAVs that would not be dependent on American patronage.[35]

Our focus is on how the American public responds to drone warfare, yet our study can inform research on other states. First, it provides a methodological model for studying attitudes toward drones. Future cross-national research may be able to ascertain the extent to which our findings reflect uniquely American concerns or whether citizens in democracies evaluate drones according to similar criteria. Second, our results suggest some of the general mechanisms that may affect attitudes toward drones in other democratic states that employ these weapons platforms. This permits some tentative generalizations in the absence of in-depth studies of other countries that use drones. To draw these generalizations we must consider how attractive pilot invulnerability and precision will be in other countries, differing levels of support for military intervention in general, and whether there are prominent strategic challenges that can serve as credible principal policy objectives.

Many of the arguments raised against drones are described as general flaws that hold true when these weapons platforms are used by any democracy. For example, Sauer and Schörnig contend that drones are attractive to democratic leaders because they make it possible to fight while minimizing financial, material, and human costs. They describe casualty aversion as "democracy-specific casualty-sensitivity" to indicate that this is a universal concern that holds true across states with this regime type and is not a unique feature of American politics.[36] The appeal of increasing selectivity with the aim of reducing civilian casualties is likewise explained as following from a general democratic appreciation of the value of life.

Horowitz and Fuhrmann offer a similar explanation of how regime type influences the decision to develop and use drones, though they show that the benefits are not unique to democracies.[37] They find a U-curve with democracies and autocracies that are on opposing ends of the spectrum of public accountability both wanting to build their drone capacities. Mixed regimes, by contrast, see little benefit in unmanned systems. Although democracies and autocracies may share an interest in drones, their incentives differ. For autocracies, drones are a means of covertly repressing domestic populations and minimizing the regime's dependence on military forces that could potentially launch a coup. Drones facilitate secrecy and the centralization of power, which are key objectives of many autocratic leaders. It is important to note that these regimes may have more accountability constraints coming from foreign countries than they do from the domestic population because of their capacity to control the flow of information internally. For democracies, the benefit of drone warfare comes chiefly from the prospect of waging foreign wars with a lower risk of upsetting public opinion and suffering electoral reversals.

We expect that democracies armed with drones will find it easier to mobilize the public in favor of military interventions, but that the overall effect of drones will have to be situated alongside other explanatory variables and that drones will probably not have a decisive influence. There are variations in countries' propensities for war. Some, like the United States, United Kingdom, Israel, and France, routinely fight small wars, while countries like Germany, Spain, and Japan tend to avoid international conflicts, despite having the ability to fight them. It is no accident that countries that fight more often and that have clearer threats have made larger investments in drone technology. The divergent attitudes seem to be relatively stable over time, reflecting the same kinds of deeper attitudes toward war that we attempted to operationalize with our military assertiveness and party affiliation variables. These variables had a consistent influence on attitudes toward war among respondents in the United States and will probably have a similar effect on citizens of other countries.

We can reach a rough estimate of how citizens in other democracies would feel about drone use by looking at a country's existing levels of support for conflicts. Opposition to war is apt to remain consistent regardless of whether drones are available. Those who offer conditional support for fighting may be swayed when drones lower the costs and raise expectations of precision. The presence of perceived threats comparable to the counterterrorism PPOs in our experiments will also be a vital consideration, with countries lacking a clear threat being less likely to pursue drones and other attack types. With this in mind, we expect that the appeal of drones will not be consistent across democracies, as other authors have predicted. Rather, the public opinion benefits should be restricted to countries with more militant publics and more plausible threats.

Nondemocracies may also find drones attractive, but they do not have the same incentives for using them as a way of escaping the constraining effects of public opinion. Sauer and Schörnig predict that nondemocracies are not as strongly affected by these considerations not only because their governments are not accountable to public opinion but also because they think that nondemocracies will generally show lower respect for human life. This seems implausible when Russia, China, and Iran are among the countries that seem to be most enthusiastic about developing their own drone forces. Sauer and Schörnig are correct in arguing that dictators may not have to worry as much about public opinion as elected officials do, but we agree with Horowitz and Fuhrmann's contention that autocracies have other incentives for developing and using drones. In particular, they need military force that can be centralized under their control and used with limited international oversight. The different incentive structure means that predicting drone use by autocracies will require a different research approach that does not rely on gauging domestic opinion. It would be more instructive to consider how much these countries depend on maintaining a good image internationally and the extent to which these leaders can entrust drone operations to reliable commanders.

Whatever their regime type, most countries will have more constraints on their use of drones than the United States will. Despite their long flight times, drones cannot operate at long ranges. They must be launched from bases that are relatively close to the target, especially if the attacker hopes to take advantage of their long loiter time. The United States has a substantial advantage over other countries with its network of military outposts around the world. Existing drones are also poorly suited to combat against other conventional militaries because they are slow and do not have good defensive systems. Finally, drone technology is imperfect.[38] The United States has lost control of aircraft and had dozens crash because of mechanical faults. Other countries may lack the resources to use such unreliable aircraft. These kinds of structural constraints are apt to limit the extent to which countries can employ drones and act as an additional constraint to prevent them from causing an increase in the incidence of wars.

Of course, there is also the question of how people view drone operations that are carried out by allied and opposing regimes. Here the goal would not be to anticipate the influence of domestic opinion on policy choices but rather to see the interaction between drone warfare and foreign perceptions. Drones could potentially strengthen or weaken relations with allied and neutral states depending on whether foreign publics see drones as being legitimate. Drones could also be used to deflect some of the costs of military intervention in terms of international public opinion.

One plausible reason why nondemocracies are interested in drones is that they will help them more effectively manage international opinion. Again, more research is needed but will only become possible over time as we gain a stronger sense of how drones are proliferating and as civilian publics learn more about how these weapons platforms are being used.

When they are positioned in the broader literature on public opinion and war, our findings show the extent to which the concerns that were previously identified as shaping attitudes toward the use of force remain salient now that drones are disrupting many of the costs and benefits identified in existing studies. Much of this work predicts that some of the most important explanations for opinions about wars—such as casualty aversion, mission objectives, and the number of civilian casualties—will be rendered largely irrelevant by these revolutionary weapons platforms. Our results reveal that many previous findings are still relevant and that drones have not influenced perceptions of conflicts enough to fundamentally alter public opinion. Instead, they show that drones (and potentially other new technologies that alter the costs of war) change the overall configuration of the factors shaping public opinion and that they are an important influence in their own right, but that the overall explanation cannot be reduced to a particular weapons system.

Appendix to Chapter 2

TABLE A2.1. Conjoint Regression Results

Model	1	2	3	4
Sample Dependent Variable:	All Choice	All Rating	Republicans Choice	Democrats Choice
Attack Type				
(Baseline: drone strike)				
Air strike	-.014	-.012	-.052	.016
	(.016)	(.018)	(.027)	(.025)
Ground troops	-.073***	-.043*	-.101***	-.091**
	(.017)	(.018)	(.028)	(.028)
Civilian Casualties				
(Baseline: low)				
Moderate	-.040*	-.048	-.027	-.052
	(.018)	(.018)**	(.033)	(.027)
High	-.159***	-.097***	-.089**	-.201***
	(.019)	(.017)	(.028)	(.026)
Mission Success				
(Baseline: low)				
Moderate	.122***	.075***	.156***	.103***
	(.018)	(.017)	(.029)	(.028)
High	.183***	.117***	.254***	.158***
	(.017)	(.019)	(.029)	(.024)
International Opinion				
(Baseline: NATO support)				
NATO opposition	-.082***	-.057*	.091*	-.098*
	(.025)	(.024)	(.039)	(.040)
Human rights group support	-.044*	-.027	-.083*	-.039
	(.023)	(.025)	(.038)	(.032)

TABLE A2.1.—*Continued*

Model	1	2	3	4
Sample Dependent Variable:	All Choice	All Rating	Republicans Choice	Democrats Choice
Security Council support	-.052* (.023)	-.028 (.024)	-.073* (.036)	-.035 (.035)
Security Council opposition	-.096*** (.024)	-.065* (.026)	-.014 (.040)	-.133*** (.034)
Principal Policy Objective				
(Baseline: counterterrorism)				
Stopping genocide	.032 (.018)	.005 (.017)	.032 (.025)	.001 (.030)
Foreign policy restraint	-.109*** (.019)	-.079*** (.017)	-.136*** (.028)	-.120*** (.032)
Target				
(Baseline: low-level enemy forces)				
Commander of enemy forces	.078*** (.013)	.029* (.013)	.122*** (.021)	.047* (.021)
Respondents	1,000	1,000	357	425
Observations	10,000	10,000	3,570	4,250

Note: * $p < .1$; ** $p < .05$; *** $p < .01$. Robust standard errors clustered on respondent in parentheses below coefficients. *Choice* is variable indicating whether attack plan is preferred over alternative; *rating* is rating of attack plan on 7-point scale.

Appendix to Chapter 3

Treatments

Each respondent was asked to read the following mock news story, which includes three treatments about the type of attack (drone, air strike, ground troops) that are randomly assigned to respondents. The text of the mock news story is below, with the text of each condition in italics:

Terrorist Camps Identified in Yemen

United States Considering Use of Military Force

Washington—Terrorists connected to al Qaeda have established bases and training camps in the country of Yemen in the Middle East. Political turmoil has prevented the government of Yemen from acting against the terrorists. Recently the Yemen branch of al Qaeda attempted to bomb an American airliner and to mail bombs to the United States.

American intelligence agencies have identified the location of the al Qaeda bases in Yemen.

[Treatment 1: The United States plans to launch attacks on these bases with missiles fired from unmanned drone aircraft to kill al Qaeda leaders and militants. The use of unmanned drones means that no American military personnel would be placed at risk.]

[Treatment 2: The United States plans to launch attacks on these bases with bombs dropped from airplanes to kill al Qaeda leaders and militants. The militants are believed to lack weapons capable of attacking these airplanes.]

[Treatment 3: The United States plans to strike the bases with American paratroopers. These American troops would attack the militants and their leaders located in the bases.]

Survey Instruments

After reading the condition described above, each respondent in both experiments answered the following questions:

On a scale from 1 to 7, where 1 indicates that the United States should definitely not carry out the attack and 7 indicates that the United States definitely should carry out the attack, how would you rate this attack plan?

- ☐ 1: Definitely do not carry out attack
- ☐ 2
- ☐ 3
- ☐ 4
- ☐ 5
- ☐ 6
- ☐ 7: Definitely carry out attack

How likely do you think it is that this attack would be to result in the death of American military personnel?

- ☐ Extremely unlikely
- ☐ Very unlikely
- ☐ Somewhat unlikely
- ☐ Neither likely nor unlikely
- ☐ Somewhat likely
- ☐ Very likely
- ☐ Extremely likely

How likely do you think it is that this attack would be to result in the death of civilians?

☐ Extremely unlikely
☐ Very unlikely
☐ Somewhat unlikely
☐ Neither likely nor unlikely
☐ Somewhat likely
☐ Very likely
☐ Extremely likely

How likely do you think it is that this attack would be to achieve its military objectives?

☐ Extremely unlikely
☐ Very unlikely
☐ Somewhat unlikely
☐ Neither likely nor unlikely
☐ Somewhat likely
☐ Very likely
☐ Extremely likely

Respondents recruited from Mturk then answered the following questions. Respondents from the YouGov panel had answered these questions earlier; their responses were provided by YouGov.

What year were you born?

What is your sex?

☐ Male
☐ Female

What racial or ethnic category best describes you?

☐ White/Caucasian
☐ Black/African American
☐ Hispanic

- ☐ Asian
- ☐ Native American
- ☐ Middle Eastern
- ☐ Mixed
- ☐ Other

What is the highest level of education you have completed?

- ☐ No high school diploma
- ☐ High school graduate
- ☐ Some college
- ☐ 2-year college degree
- ☐ 4-year college degree
- ☐ Graduate school

How would you describe your marital status?

- ☐ Married
- ☐ Separated
- ☐ Divorced
- ☐ Widowed
- ☐ Single
- ☐ Domestic partnership

How would you describe your family's total income for last year?

- ☐ Less than $10,000
- ☐ $10,000–$19,999
- ☐ $20,000–$29,999
- ☐ $30,000–$39,999
- ☐ $40,000–$49,999
- ☐ $50,000–$59,999
- ☐ $60,000–$69,999
- ☐ $70,000–$79,999
- ☐ $80,000–$89,999
- ☐ $90,000–$99,999
- ☐ $100,000–$119,999
- ☐ $120,000–$149,999

- ☐ $150,000–$199,999
- ☐ $200,000–$249,999
- ☐ $250,000–$349,000
- ☐ $350,000 or more

How would you describe your political views?

- ☐ Very liberal
- ☐ Liberal
- ☐ Moderate
- ☐ Conservative
- ☐ Very conservative

Generally speaking, do you usually think of yourself as a Republican, a Democrat, an Independent, or something else?

- ☐ Strong Democrat
- ☐ Not very strong Democrat
- ☐ Lean Democrat
- ☐ Independent
- ☐ Lean Republican
- ☐ Not very strong Republican
- ☐ Strong Republican

How would you describe your religious beliefs?

- ☐ Protestant
- ☐ Roman Catholic
- ☐ Mormon
- ☐ Eastern or Greek Orthodox
- ☐ Jewish
- ☐ Muslim
- ☐ Buddhist
- ☐ Hindu
- ☐ Atheist
- ☐ Agnostic
- ☐ Nothing in particular
- ☐ Something else

Demographics

TABLE A3.1. Demographic Characteristics of Respondents

	YouGov	Mechanical Turk	ANES
Age	48	33	38–53
Income	$50,000–$59,000	$50,000–$59,000	$60,000–$74,999
Ideology	Moderate	Liberal	Moderate
Gender			
Male	46	52	48
Female	54	48	52
Education			
Less than a high school degree	6	1	2
High school degree	35	11	38
Some college/2 year degree	33	36	30
College degree/post-grad	26	52	29
Race and ethnicity			
White or Caucasian	72	82	71
Black non-Hispanic	10	6	12
Latino or Hispanic	10	4	11
Other	16	8	6
Party identification			
Democrat	35	52	46
Independent	29	26	14
Republican	25	21	39
Religion			
Atheist or agnostic	12	38	13
Marital status			
Married	51	43	48
Observations	1,000	1,205	

Note: Age, income, and ideology are median values; remaining variables are percentages. ANES is the American National Election Studies; data is from the 2012 survey reported in American National Election Studies, *User's Guide and Codebook for the ANES 2012 Time Series Study*, the University of Michigan and Stanford University, May 28, 2015, available at http://electionstudies.org/studypages/anes_timeseries_2012/anes_timeseries_2012_userguidecodebook.pdf

Appendix to Chapter 4

Experimental Treatments

Treatment: Counterterrorism and Drone Strikes

Terrorist Camps Identified in Yemen

United States Considering Use of Military Force

Washington (AP)—Terrorists connected to al Qaeda have established bases and training camps in the country of Yemen in the Middle East. Political turmoil has prevented the government of Yemen from acting against the terrorists. Recently the Yemen branch of al Qaeda attempted to bomb an American airliner and to mail bombs to the United States. American intelligence agencies have identified the location of the al Qaeda bases in Yemen. The United States plans to launch attacks on these bases with missiles fired from unmanned drone aircraft to kill al Qaeda leaders and militants. The use of unmanned drones means that no American military personnel would be placed at risk.

Treatment: Counterterrorism and Air Strikes

Terrorist Camps Identified in Yemen

United States Considering Use of Military Force

Washington (AP)—Terrorists connected to al Qaeda have established bases and training camps in the country of Yemen in the Middle East. Political turmoil has prevented the government of Yemen from acting against the terrorists. Recently the Yemen branch of al Qaeda attempted to bomb an American airliner and to mail bombs to the United States. American intelligence agencies have identified the location of the al Qaeda bases in Yemen. The United States plans to launch attacks on these bases with bombs dropped from airplanes to kill al Qaeda leaders and militants. The militants are believed to lack weapons capable of attacking these airplanes.

Treatment: Counterterrorism and Ground Troops

Terrorist Camps Identified in Yemen

United States Considering Use of Military Force

Washington (AP)—Terrorists connected to al Qaeda have established bases and training camps in the country of Yemen in the Middle East. Political turmoil has prevented the government of Yemen from acting against the terrorists. Recently the Yemen branch of al Qaeda attempted to bomb an American airliner and to mail bombs to the United States. American intelligence agencies have identified the location of the al Qaeda bases in Yemen. The United States plans to strike the bases with American paratroopers. These American troops would attack the militants and their leaders located in the bases.

Treatment: Foreign Policy Restraint and Drone Strikes

Yemen Threatens International Oil Supply

United States Considering Use of Military Force

Washington (AP)—The country of Yemen has threatened to attack ships carrying oil from the Persian Gulf. Yemen's location in the Middle East places its military's missiles in range of oil tankers that transport petroleum to international markets. The threat has led to sharp increases in oil prices and uncertainty in world markets about the future supply of petroleum. American intelligence agencies have identified the location of the Yemen military bases that can attack the oil shipments. The United States plans to launch attacks on these bases with missiles fired from unmanned drone aircraft. The use of unmanned drones means that no American military personnel would be placed at risk.

Treatment: Foreign Policy Restraint and Air Strikes

Yemen Threatens International Oil Supply

United States Considering Use of Military Force

Washington (AP)—The country of Yemen has threatened to attack ships carrying oil from the Persian Gulf. Yemen's location in the Middle East places its military's missiles in range of oil tankers that transport petroleum to international markets. The threat has led to sharp increases in oil prices and uncertainty in world markets about the future supply of petroleum. American intelligence agencies have identified the location of the Yemen military bases that can attack the oil shipments. The United States plans to launch attacks on these bases with bombs dropped from airplanes. The Yemen military forces at these bases are believed to lack weapons capable of attacking these airplanes.

Treatment: Foreign Policy Restraint and Ground Troops

Yemen Threatens International Oil Supply

United States Considering Use of Military Force

Washington (AP)—The country of Yemen has threatened to attack ships carrying oil from the Persian Gulf. Yemen's location in the Middle East places its military's missiles in range of oil tankers that transport petroleum to international markets. The threat has led to sharp increases in oil prices and uncertainty in world markets about the future supply of petroleum. American intelligence agencies have identified the location of the Yemen military bases that can attack the oil shipments. The United States plans to strike the bases with American paratroopers. These American troops would attack Yemeni military forces at the bases.

Treatment: Humanitarian Intervention and Drone Strikes

Government of Yemen Accused of Genocide

United States Considering Use of Military Force

Washington (AP)—Military forces in the country of Yemen are committing atrocities. Members of minority ethnic groups as well as opponents of the country's government are subject to mass killings, arrest, and torture. International human rights groups are calling these actions genocide and demand that the international community intervene. American intelligence agencies have identified the location of the Yemen military bases involved in the genocide. The United States plans to launch attacks on these bases with missiles fired from unmanned drone aircraft. The use of unmanned drones means that no American military personnel would be placed at risk.

Treatment: Humanitarian Intervention and Air Strikes

Government of Yemen Accused of Genocide

United States Considering Use of Military Force

Washington (AP)—Military forces in the country of Yemen are committing atrocities. Members of minority ethnic groups as well as opponents of the country's government are subject to mass killings, arrest, and torture. International human rights groups are calling these actions genocide and demand that the international community intervene. American intelligence agencies have identified the location of the Yemen military bases involved in the genocide. The United States plans to launch attacks on these bases with bombs dropped from airplanes. The Yemen military forces at these bases are believed to lack weapons capable of attacking these airplanes.

Treatment: Humanitarian Intervention and Ground Troops

Government of Yemen Accused of Genocide

United States Considering Use of Military Force

Washington (AP)—Military forces in the country of Yemen are committing atrocities. Members of minority ethnic groups as well as opponents of the country's government are subject to mass killings, arrest, and torture. International human rights groups are calling these actions genocide and demand that the international community intervene. American intelligence agencies have identified the location of the Yemen military bases involved in the genocide. The United States plans to strike the bases with American paratroopers. These American troops would attack Yemeni military forces at the bases.

Treatment: Internal Political Change and Drone Strikes

Government of Yemen May Collapse

United States Considering Use of Military Force

Washington (AP)—The government of Yemen, an ally of the United States, is on the verge being overthrown. Rebel groups in the north of the country have defeated government forces in a series of battles and are approaching the capital. The government has appealed to the United States to provide military support. American intelligence agencies have identified the location of the rebel bases in Yemen. The United States plans to launch attacks on these bases with missiles fired from unmanned drone aircraft to kill rebel leaders and militants. The use of unmanned drones means that no American military personnel would be placed at risk.

Treatment: Internal Political Change and Air Strikes

Government of Yemen May Collapse

United States Considering Use of Military Force

Washington (AP)—The government of Yemen, an ally of the United States, is on the verge being overthrown. Rebel groups in the north of the country have defeated government forces in a series of battles and are approaching the capital. The government has appealed to the United States to provide military support. American intelligence agencies have identified the location of the rebel bases in Yemen. The United States plans to launch attacks on these bases with bombs dropped from airplanes to kill rebel leaders and militants. The militants are believed to lack weapons capable of attacking these airplanes.

Treatment: Internal Political Change and Ground Troops

Government of Yemen May Collapse

United States Considering Use of Military Force

Washington (AP)—The government of Yemen, an ally of the United States, is on the verge being overthrown. Rebel groups in the north of the country have defeated government forces in a series of battles and are approaching the capital. The government has appealed to the United States to provide military support. American intelligence agencies have identified the location of the rebel bases in Yemen. The United States plans to strike the bases with American paratroopers. These American troops would attack the rebels and their leaders located in the bases.

Survey Instrument

How do you feel about this attack plan?

- ☐ I strongly approve of this attack plan. (1)
- ☐ I somewhat approve of this attack plan. (2)
- ☐ I somewhat disapprove of this attack plan. (3)
- ☐ I strongly disapprove of this attack plan. (4)

How likely is it that the attack plan will succeed?

- ☐ Very likely (1)
- ☐ Somewhat likely (2)
- ☐ Somewhat unlikely (3)
- ☐ Very unlikely (4)

What is your best guess of the number of American soldiers that would be killed or injured if the attack were carried out?

- ☐ None (1)
- ☐ Between 1 and 10 soldiers (2)
- ☐ Between 11 and 100 soldiers (3)
- ☐ More than 100 soldiers (4)

The United States faces many challenges today, at home as well as overseas. Among these challenges, how important do you think it is to address the problems discussed in the news story you read earlier?

- ☐ Very important (1)
- ☐ Somewhat important (2)
- ☐ Neither important nor unimportant (3)
- ☐ Somewhat unimportant (4)
- ☐ Not very important (5)

Respondents then answered questions about their education level, age, income, ethnicity, gender, political ideology, and party identification; the survey items for these questions are identical to those used in the convenience sample reported in chapter 3.

Demographics

TABLE A4.1. Demographic Characteristics of Respondents

	PPO Experiment	ANES
Age	29	38–53
Ideology	Moderate	Moderate
Gender		
Male	56	48
Female	44	52
Education		
Less than a high school degree	0	2
High school degree	10	38
Some college/2 year degree	37	30
College degree/post-grad	52	29
Race and ethnicity		
White or Caucasian	76	71
Black non-Hispanic	6	12
Latino or Hispanic	5	11
Other	12	6
Party identification		
Democrat	55	46
Independent	21	14
Republican	24	39
Observations	3,700	

Note: Age, income, and ideology are median values; remaining variables are percentages. ANES is the American National Election Studies; data is from the 2012 survey reported in American National Election Studies, *User's Guide and Codebook for the ANES 2012 Time Series Study*, the University of Michigan and Stanford University, May 28, 2015, available at http://electionstudies.org/studypages/anes_timeseries_2012/anes_timeseries_2012_userguidecodebook.pdf

Appendix to Chapter 5

Treatments

Respondents first read the text below, which randomly varies the attack type and alternatives to military action:

U.S. Plans Attack on Insurgents in Iraq

Washington (AP)—U.S. intelligence agencies have identified the location of an insurgent compound in Iraq. The insurgents are planning to attack Iraqi soldiers guarding a bridge over a wide river. If the insurgents capture the bridge, they could move their forces into position to threaten a major Iraqi city.

The U.S. is planning to attack the insurgents [*drone strike treatment*: with guided missiles launched from drones—remotely piloted aircraft without an on-board crew / *air strike treatment:* with guided missiles launched from strike aircraft, each of which has an on-board crew of two officers / *ground troops treatment:* with a unit of American ground troops, who would approach the compound and fire guided missiles from about 25 yards away].

The Secretary of Defense has authorized the attack. He was overheard saying that this decision was an easy one because the alternatives to American intervention with [*drone strike treatment*: drone strikes / *air strike treatment:* air strikes / *ground troops treatment:* ground troops], such as dropping weapons to the Iraqi troops or airlifting more Iraqi troops to defend the bridge, were [*bad options:* unlikely / *good options:* likely] to work.

Survey Instrument

How do you feel about this attack plan by the United States?

- ☐ I strongly approve of this attack plan.
- ☐ I somewhat approve of this attack plan.
- ☐ I somewhat disapprove of this attack plan.
- ☐ I strongly disapprove of this attack plan.

Do you think it is a good idea or bad idea for the United States to carry out this attack plan?

- ☐ A very good idea
- ☐ A somewhat good idea
- ☐ A somewhat bad idea
- ☐ A very bad idea

Would you say that you favor or oppose the United States' plan to launch this attack?

- ☐ I strongly favor this attack plan
- ☐ I somewhat favor this attack plan
- ☐ I somewhat oppose this attack plan
- ☐ I strongly oppose this attack plan

Based on the information in this news story, the Secretary of Defense thought his decision was a difficult one.

- ☐ True
- ☐ False

Based on the information in this news story, the alternatives to an American attack, such as dropping weapons to the Iraqi troops or airlifting more Iraqi troops to defend the bridge, were likely to work.

- ☐ True
- ☐ False

Based on the information in this news story, how confident are you that the plan to attack the insurgent compound will be successful?

- ☐ Extremely confident
- ☐ Somewhat confident
- ☐ Not very confident
- ☐ I am confident it will NOT succeed

Regardless of whether you support the American attack, how many deaths would you expect the U.S. military to suffer in this operation?

- ☐ Zero deaths
- ☐ Between 1 and 10 deaths
- ☐ Between 11 and 50 deaths
- ☐ More than 50 deaths

Next we have some questions about your background.

What is the highest grade of school or year of college you have completed?

- ☐ No high school diploma
- ☐ High school diploma
- ☐ Some college, no bachelor's degree
- ☐ Bachelor's degree
- ☐ Graduate degree

What year were you born?

What racial or ethnic category best describes you?

- ☐ White/Caucasian
- ☐ African American
- ☐ Hispanic
- ☐ Asian

☐ Native American
☐ Pacific Islander
☐ Other or multiple categories

What is your sex?

☐ Male
☐ Female

How would you describe your political views?

☐ Very conservative
☐ Conservative
☐ Moderate
☐ Liberal
☐ Very liberal

Generally speaking, do you usually think of yourself as a Republican, a Democrat, an independent, or something else?

☐ Republican
☐ Democrat
☐ Independent
☐ Something else

(Asked if respondent identifies as Democrat) Would you call yourself a strong Democrat or a not very strong Democrat?

☐ Strong Democrat
☐ Not very strong Democrat

(Asked if respondent identifies as Republican) Would you call yourself a strong Republican or a not very strong Republican?

☐ Strong Republican
☐ Not very strong Republican

(Asked if respondent identifies as Independent) Do you think of yourself as closer to the Republican or Democratic Party?

- ☐ Republican Party
- ☐ Democratic Party
- ☐ Just independent

Demographics

TABLE A5.1. Demographic Characteristics of Respondents

	Moral Hazard Experiment	ANES
Age	30	38–53
Ideology	Moderate	Moderate
Gender		
Male	59	48
Female	41	52
Education		
Less than a high school degree	1	2
High school degree	10	38
Some college/2 year degree	34	30
College degree/post-grad	52	29
Race and ethnicity		
White or Caucasian	71	71
Black non-Hispanic	11	12
Latino or Hispanic	7	11
Other	11	6
Party Identification		
Democrat	60	46
Independent	17	14
Republican	23	39
Observations	1,241	

Note: Age, income, and ideology are median values; remaining variables are percentages. ANES is the American National Election Studies; data is from the 2012 survey reported in American National Election Studies, *User's Guide and Codebook for the ANES 2012 Time Series Study*, the University of Michigan and Stanford University, May 28, 2015, available at http://electionstudies.org/studypages/anes_timeseries_2012/anes_timeseries_2012_userguidecodebook.pdf

Appendix to Chapter 6

This appendix provides additional details about the two experiments discussed in chapter 6. These include the texts of the treatments for the experiments, details regarding administration of the survey experiments and demographic information about the respondents, additional information regarding difference of means tests, and regression results.

Treatments for First Experiment

All Treatments: U.S. Military Plans Strike Against Insurgent Hideouts

Insurgents have established hideouts in the country of Pakistan. These insurgents have attacked American military forces based in neighboring Afghanistan.

American intelligence agencies have identified the location of the insurgents' hideouts.

High selectivity/unmanned: The United States plans to launch attacks on these hideouts with missiles fired from unmanned "drone" aircraft. The use of unmanned drones means that no American military personnel would be placed at risk.

All other treatments: The United States plans to launch attacks on these hideouts with (**high selectivity, manned:** missiles fired/ **moderate and low selectivity:** bombs dropped) from airplanes.

The insurgents do not have weapons capable of attacking these airplanes. This means that no Americans would be placed at risk.

The insurgents' hideouts are in farming villages near the houses of civilians. These civilians have no connection to the insurgent group. It is possible that the drone strikes would accidentally kill civilians.

High selectivity, unmanned: But experts conclude the chance of killing civilians is very low. The drones are equipped with high-definition video cameras, so strikes can be launched when no civilians are nearby. The missiles fired by drones are small and accurate enough to be aimed at specific room in the hideouts. This makes it less likely that civilians would be harmed. Recent drone strikes have resulted in very few civilian casualties. In 2012, for example, only 3 percent of those killed in drone strikes were civilians.

High selectivity, manned: But experts conclude the chance of killing civilians is very low. The airplanes are equipped with high-definition video cameras, so strikes can be launched when no civilians are nearby. The missiles are small and accurate enough to be aimed at specific rooms in the hideouts. This makes it less likely that civilians would be harmed. Recent missile strikes have resulted in very few civilian casualties. In 2012, for example, only 3 percent of those killed in drone strikes were civilians.

Moderate selectivity: Experts conclude the attacks could kill civilians. The bombs might drift off course and miss their target. The pilots will not be able to see if civilians are near the target when they drop the bombs. Even if the bombs hit the hideouts, the explosions they create might injure nearby civilians. Past bomb attacks have killed civilians by accident. In 2012, for example, 35 percent of those killed in bomb attacks were civilians.

Low selectivity: Experts conclude the attacks could kill civilians. The bombs might drift off course and miss their target. The pilots will not be able to see if civilians are located near the target when they drop the bombs. Even if the bombs hit the hideouts, the explosions they create might injure nearby civilians. Past bomb attacks have killed civilians by accident. In 2012, for example, fully 50 percent of those killed in bomb attacks were civilians.

Treatment for Second Experiment

High-Threat Treatments: U.S. Military Plans Strike Against Terrorist Bases in Yemen

Washington—Insurgents have established bases in the country of Yemen. Political turmoil has prevented the government of Yemen from acting against this insurgent group. Recently the group has attempted to bomb an American airliner and to mail bombs to the United States.

Low-Threat Treatments: U.S. Military Plans Strike Against Terrorist Bases in Yemen

Washington—Insurgents have established bases in the country of Yemen. Political turmoil has prevented the government of Yemen from acting against this insurgent group. Recently the group has attacked military and civilian targets in Yemen. Yemen is an important ally of United States, but the insurgents have not threatened or attacked the United States.

All Treatments: The United States plans to try to kill insurgents by launching attacks on the bases with missiles fired from unmanned drone aircraft. The use of unmanned drones means that no American military personnel would be placed at risk.

High-Distinction Treatments: American intelligence agencies have identified members of the insurgent group and the location of their bases. Drones equipped with video cameras allow them to monitor the bases closely. They are confident they can launch missile strikes when the insurgents are at the bases and no civilians are nearby.

Low-Distinction Treatments: American intelligence agencies have located the insurgents' bases, but have had difficulty identifying members of the insurgent group. Drones equipped with video cameras allow them to monitor the bases closely. They plan to launch drone strikes when people arrive at the bases who appear to be insurgents, such as men of military age. American officials acknowledge that some of those killed might not be active members of the insurgent group.

Violates International Law Treatments: The drone strikes would violate international law. The United States has signed treaties that allow the use of deadly force only under certain circumstances. Most experts in international law conclude that the drone strikes would violate these treaties.

May Violate International Law Treatments: The drone strikes might violate international law. The United States has signed treaties that allow the use of deadly force only under certain circumstances. But experts do not agree if the drone strikes would violate these treaties, as there is much debate about what they really mean.

Survey Instrument for First Experiment

What is your gender?

- ☐ Female
- ☐ Male

What racial or ethnic group best describes you?

- ☐ White or Caucasian
- ☐ Black or African American
- ☐ Latino or Hispanic
- ☐ Asian
- ☐ Native American
- ☐ Other

What is the highest grade of school or year of college you have completed?

- ☐ No high school diploma
- ☐ High school diploma
- ☐ Some college, no bachelor's degree
- ☐ Bachelor's degree
- ☐ Graduate degree

What year were you born?

Please pick the category of the income group that includes the income of all members of your family living with you in 2011 before taxes. This figure should include salaries, wages, pensions, dividends, interest, and all other income.

- ☐ Less than $25,000
- ☐ $25,000 to $49,999
- ☐ $50,000 to $79,999
- ☐ $80,000 to $119,999
- ☐ $120,000 and above

Next are some questions to help us see how much information about politics gets out to the public. Please answer these questions on your own, without asking anyone or looking up the answers. Many people don't know the answers to these questions, but we would be grateful if you would please answer every question, even if you are not sure what the right answer is.

Do you happen to know how many times an individual can be elected President of the United States under current laws?

- ☐ 1 time
- ☐ 2 times
- ☐ 3 times
- ☐ No limit

How many U.S. Senators are there from each state?

- ☐ One
- ☐ Two
- ☐ Three
- ☐ Four

For how many years is a member of the United States House of Representatives elected? That is, how many years are there in one full term of office for a U.S. House member?

- ☐ 2 years
- ☐ 4 years
- ☐ 6 years
- ☐ 8 years

What percentage vote of the House and the Senate is needed to override a Presidential veto?

- ☐ A bare majority
- ☐ Two-thirds
- ☐ Three-fourths
- ☐ Ninety percent

According to federal law, if the president of the United States dies, is no longer willing or able to serve, or is removed from office by Congress, the vice president would become the president. If the vice president were unable or unwilling to serve, who would be eligible to become president next?

- ☐ The chief justice of the supreme court
- ☐ The secretary of state
- ☐ The speaker of the House of Representatives

Please indicate how confident you are in the US military:

- ☐ Very confident
- ☐ Somewhat confident
- ☐ Not very confident
- ☐ Not at all confident

Generally speaking, do you usually think of yourself as a Democrat, a Republican, an independent, or what?

- ☐ Strong Democrat
- ☐ Democrat
- ☐ Independent
- ☐ Republican
- ☐ Strong Republican

How would you describe your political views?

- ☐ Very conservative
- ☐ Conservative
- ☐ Moderate
- ☐ Liberal
- ☐ Very liberal

Have you or has anyone in your household ever served in the US military or the military reserves?

- ☐ Yes
- ☐ No

All respondents answered the following questions after reading the treatment:

How do you feel about this plan to attack insurgent bases?

- ☐ I strongly approve of this attack plan.
- ☐ I somewhat approve of this attack plan.
- ☐ I somewhat disapprove of this attack plan.
- ☐ I strongly disapprove of this attack plan.

If you had to choose, would you support this plan to attack insurgent bases?

- ☐ Yes, I would support this attack plan.
- ☐ No, I would not support this attack plan.

If this attack is carried out, do you think it will cause civilian casualties?

- ☐ Definitely
- ☐ Very Probably
- ☐ Probably
- ☐ Possibly
- ☐ Probably Not
- ☐ Very Probably Not

Overall, taking into account the risks and benefits, how satisfied are you with this attack?

- ☐ Very satisfied
- ☐ Somewhat satisfied
- ☐ Neither satisfied nor dissatisfied
- ☐ Somewhat dissatisfied
- ☐ Very dissatisfied

According to the news story you read earlier, what percentage of those killed in similar attacks in 2012 were civilians?

- ☐ None
- ☐ 3 percent
- ☐ 35 percent
- ☐ 50 percent
- ☐ 85 percent

Respondents in all treatments then read the following statement:

The military carried out the attack described on the previous page. The attack did kill the leaders of the insurgent group. However, the strike also killed nearby civilians unconnected to the insurgent group.

Now that you know the results of the attack, please answer the final set of questions on the following page.

Overall, taking into account the risks and benefits, how satisfied are you with this attack?

- ☐ Very satisfied
- ☐ Somewhat satisfied
- ☐ Neither satisfied nor dissatisfied
- ☐ Somewhat dissatisfied
- ☐ Very dissatisfied

In past conflicts that have accidentally killed civilians, the United States has offered compensation to the victims' families. How much compensation should it offer in this case to each civilian victim's family? The average amount suggested by people like you is $20,000. Please select the amount you think should be offered from the drop-down menu.

- ☐ No compensation
- ☐ $5,000
- ☐ $10,000
- ☐ $15,000
- ☐ $20,000
- ☐ $25,000
- ☐ $30,000
- ☐ $35,000
- ☐ $40,000 or more

How surprised were you to learn that civilians were accidentally killed in this attack?

- ☐ Very surprised
- ☐ Somewhat surprised
- ☐ A little surprised
- ☐ Not very surprised
- ☐ Not at all surprised

How important do you think it is that the United States issue a public apology for accidentally killing civilians in this attack?

- ☐ Very important
- ☐ Important
- ☐ Moderately important
- ☐ Of little importance
- ☐ Not at all important

According to the information you read, were civilians killed in this attack?

- ☐ No
- ☐ Yes

Survey Instrument for Second Experiment

Respondents answered the same questions used in the first experiment in this chapter measuring gender, ethnicity, education, income, age, military service, and attention to politics, confidence in the military, party identification, and political ideology.

Respondents were randomly assigned to one of twelve treatment groups and presented with the appropriate vignette. They then answered the following two questions:

How do you feel about this plan to attack insurgent bases?

- ☐ I strongly approve of this attack plan.
- ☐ I somewhat approve of this attack plan.
- ☐ I somewhat disapprove of this attack plan.
- ☐ I strongly disapprove of this attack plan.

If you had to choose, would you support this plan to attack insurgent bases?

- ☐ Yes, I would support this attack plan.
- ☐ No, I would not support this attack plan.

Demographics

TABLE A6.1. Demographic Characteristics of Respondents

	First Experiment	Second Experiment	Population
Gender			
Male	62	69	49
Female	38	31	51
Education			
Less than a high school degree	1	1	13
High school degree	13	9	30
Some college	41	46	29
4-year college degree	34	35	18
Graduate degree	11	9	10
Race and ethnicity			
White or Caucasian	79	77	74
Black or African American	5	5	13
Latino or Hispanic	4	5	
Asian	10	11	5
Native American	0	1	1
Other	2	2	16
Party identification			
Strong Democrat	11	12	19
Democrat	35	35	32
Independent	39	39	11
Republican	11	11	25
Strong Republican	3	2	13
Observations	731	1,713	

Note: Data are percentages. Population is the United States; data from U.S. Census Bureau and American National Election Survey. Note that race and ethnicity sums to greater than 100 percent in the population column because respondents could choose more than one race.

Ordered Logistic Regression Analysis for First Experiment

The use of difference of means tests to this point in the analysis may be problematic for two reasons. First, it assumes that the difference between the possible answer choices on ordinal scales reflect identical differences in the range of the respondent's preferences. This might not be the case. Second, although the treatments are balanced in terms of the demographic characteristics of the respondents, it is possible that other individual-level characteristics are driving the results. To address these concerns, table A6.2 below reports the results of ordered logistic regression models for the dependent variables in the first experiment. The excluded category is the unmanned high-precision treatment, and the model includes dummy variables measuring assignment to the moderate- or low-precision treatments. Respondents assigned to the manned, high-precision treatment are not included in the model. The dependent variable in the first model is the amount of compensation the respondent would be willing to have the United States government offer to the family of each civilian victim. This is a 9-point ordinal measure ranging in $5,000 increments from no compensation to $40,000 or more in compensation. The dependent variable in the second model is the 5-point scale measuring the degree to which the respondent thinks it is important to offer an apology for the attack. In the third model, the dependent variable is the degree of satisfaction with the attack before the respondent learns that it resulted in civilian casualties minus the response to the same question after learning about civilian casualties. The results of all three models are similar to those from the difference of means analysis reported in the chapter. Assignment to the moderate- or low-precision treatment reduces compensation, compared to the excluded treatment. Neither of these treatments influence the degree to which the respondent thinks it is important to issue an apology. Assignment to the moderate- or low-precision treatment leads to a smaller amount of satisfaction change compared to those in the excluded treatment. Here few of the control variables are statistically significant, although confidence in the military reduces the amount of compensation the respondent supports and the importance attached to an apology. Table A6.3 reports results of the same models, but using the manned, precision treatment as the excluded category and not including respondents assigned to the unmanned, precision treatments. This means that the coefficients on the moderate-precision and low-precision variables estimate the difference in the dependent variable between respondents assigned to either of these treatments and the excluded category. Consistent with the difference of means discussion in the chapter, we see that these differences are not significantly significant for compensation but that respondents assigned to the low-precision or moderate-precision treatments were more likely to approve of offering an apology for civilian casualties.

TABLE A6.2. Ordered Logistic Regression Results for First Experiment

Dependent Variable	Compensation	Apology	Satisfaction Change
Moderate precision	–.38*	.20	1.66*
	(.19)	(.20)	(.22)
Low precision	–.39*	.17	1.67*
	(.20)	(.20)	(.22)
Confidence in military	.30*	–.43*	–.03
	(.11)	(.11)	(.11)
Partisanship	.18*	.16*	.04
	(.09)	(.09)	(.10)
Gender	.32	.16	–.02
	(.17)	(.18)	(.18)
Age	.00	.01	.01
	(.01)	(.01)	(.01)
Education	–.09	.19*	–.12
	(.09)	(.10)	(.10)
Caucasian	–.22	.07	.16
	(.19)	(.20)	(.21)
Income	–.20	–.01	–.06
	(.07)	(.07)	(.07)
AIC	1,912	1,411	1,296

Note: Unmanned, high-precision is the excluded category. Standard errors in parentheses; * $p < .05$.

TABLE A6.3. Ordered Logistic Regression Results for First Experiment

Dependent Variable	Compensation	Apology	Satisfaction Change
Moderate precision	−.03	.41*	1.64*
	(.19)	(.20)	(.22)
Low precision	.03	.36	1.64*
	(.19)	(.21)	(.23)
Confidence in military	.25*	−.11	.17
	(.11)	(.11)	(.11)
Partisanship	−.20*	.19*	.06
	(.08)	(.08)	(.09)
Gender	.32*	.59*	−.05
	(.16)	(.17)	(.18)
Age	−.00	−.00	.00
	(.00)	(.00)	(.00)
Education	−.13	.18	−.11
	(.09)	(.09)	(.10)
Caucasian	−.23	.16	.07
	(.20)	(.20)	(.22)
Income	−.03	.11	.01
	(.07)	(.07)	(.07)
AIC	2,015	1,410	1,298
Observations	553	553	553

Note: Manned, high-precision is the excluded category. Standard errors in parentheses; * $p < .05$.

Ordered Logistic Regression Analysis for Second Experiment

We also carried out regression analysis for the second experiment. Table A6.4 reports the results of two ordered logistic regression models that correspond to the difference of means analysis summarized in the chapter. These models include only respondents in the four treatment groups that make no mention of international law. The dependent variable in both is the respondent's answer to the four-point question about support for the use of force. Dummy variables were used to capture assignment to each condition, and the high-proportionality, high-distinction treatment is the excluded category. Control variables include confidence in the military, military experience, party identification, gender, age, education, race, and income.[1]

Regression analysis is particularly useful here because of the potentially confounding effect of confidence in the military. Larson and Savych hold that civilian casualties do not systematically influence attitudes toward the use of force because most Americans believe that the military will take reasonable steps to avoid killing civilians.[2] The treatment conditions used here directly measure some decisions that could be made to avoid civilian deaths. Proportionality measures the extent to which the degree of threat influences decisions to use force that might risk civilian casualties, while distinction captures how willing the respondent is to risk killing civilians in order to achieve a military objective.

Consistent with the findings reported in the chapter, assignment to a condition other than the baseline of high proportionality, high distinction results in statistically significant decreases in support for the use of force. Confidence in the military has a negative and significant influence on support, indicating that those with less confidence in the military are less likely to agree to the decision to launch attacks. Military experience, identification as a Republican, and income are associated with greater support for the use of force, and a more liberal political ideology reduces support for the use of force. Note as well that the inclusion of control variables does not result in substantial changes in the coefficients for the experimental treatments.

We used a similar approach to test the robustness of the difference of means reported in figure 6.1; these models are presented below in tables A6.5, A6.6, and A6.7. Table A6.5 uses respondents assigned to the same treatments summarized in the first panel of figure 6.1. The models differ in terms of the treatment groups that they include. Respondents in the first model include those assigned to the high-proportionality, high-distinction treatment that makes no mention of international law and the high-proportionality, high-distinction treatment which states that the attacks violate international law. Respondents in subsequent models include those whose vignettes did and did not mention international law and the treatments described in the first row.

We created a dummy variable that takes a value of 1 if the respondent read that the attack violates international law, and a value of zero otherwise. This is the first independent variable listed in the table, and it tests if exposure to this information influences support for the attack. All the coefficients are negative and statistically significant, which is consistent with the difference of means reported in the chapter.

Tables A6.6 and A6.7 perform the same exercise for the center and lower panels of figure 6.1. In table A6.6, respondents read vignettes that made no mention of international law, or stated that the attacks might violate the law. In table A6.7, respondents read vignettes that stated that the attack might violate international law, or that it would violate international law. The dummy variable measuring assignment to these treatments is significant and negative for the same comparisons in which this was the case for the difference of means tests. In addition, the first model in table A6.7 suggests that definitively violating international law reduces support compared to treatments in which the attack is described as possibly violating the law when the degree of proportionality and distinction are both high.

Turning to the control variables, few of these have a consistent statistically significant relationship with the dependent variable across the models. The exception to this is again confidence in the military; in all the models, respondents with less confidence in the military were less likely to approve of the attack plan described in their assigned treatment.

TABLE A6.4. Ordered Logistic Regression Results for Distinction and Threat Treatments in Second Experiment

	Model 1	Model 2
Low threat, high distinction	-.56*	-.77*
	(.22)	(.23)
High threat, low distinction	-1.26	-1.32*
	(.23)*	(.24)
Low threat, low distinction	-1.59*	-1.76*
	(.23)	(.24)
Confidence in military		-.82*
		(.11)
Military experience		.78*
		(.21)
Partisanship		.17*
		(.08)
Gender		.33*
		(.17)
Age		.01
		(.01)
Education		-.01
		(.10)
Caucasian		.03
		(.19)
Income		.15*
		(.07)
AIC	1,401	1,306
Observations	572	571

Note: Standard errors in parentheses; * $p < .05$.

TABLE A6.5. Effect of Violating International Law Compared to No Mention of International Law in Second Experiment

	Model 3 High Proportionality, High Distinction	Model 4 Low Proportionality, High Distinction	Model 5 High Proportionality, Low Distinction	Model 6 Low Proportionality, Low Distinction
Violates international law	-1.20*	-1.10*	-1.10*	-.64*
	(.24)	(.23)	(.24)	(.22)
Confidence in military	-.76*	-.74*	-.72*	-.68*
	(.18)	(.16)	(.16)	(.15)
Military experience	.70*	.07	.32	.56
	(.31)	(.30)	(.48)	(.30)
Partisanship	.22	.35*	.28*	.16
	(.12)	(.11)	(.13)	(.12)
Gender	.71*	.51*	.06	.15
	(.25)	(.24)	(.26)	(.25)
Age	.02	-.01	.01	.01
	(.01)	(.01)	(.01)	(.01)
Education	-.04	.15	.11	-.15
	(.15)	(.14)	(.15)	(.14)
Caucasian	-.22	-.04	-.04	.04
	(.25)	(.27)	(.26)	(.26)
Income	.01	.04	.30	.18
	(.10)	(.09)	(.09)	(.09)
AIC	671	676	607	669
Observations	282	290	276	290

Note: Standard errors in parentheses; * $p < .05$.

TABLE A6.6. Effect of May Violate International Law Compared to No Mention of International Law in Second Experiment

	Model 7 High Proportionality, High Distinction	Model 8 Low Proportionality, High Distinction	Model 9 High Proportionality, Low Distinction	Model 10 Low Proportionality, Low Distinction
May violate international law	–.76*	–.74*	–.04	–.04
	(.23)	(.23)	(.22)	(.22)
Confidence in military	–.79*	–.97*	–.81*	–.87*
	(.16)	(.16)	(.15)	(.16)
Military experience	.66*	.44*	.21	.32
	(.31)	(.29)	(.28)	(.28)
Partisanship	.20	.23*	–.04	.15
	(.12)	(.11)	(.13)	(.13)
Gender	.73*	.25*	.30	.07
	(.25)	(.25)	(.25)	(.25)
Age	.02	.01	.02	.00
	(.01)	(.01)	(.01)	(.01)
Education	–.06	.30*	–.01	–.17
	(.14)	(.15)	(.14)	(.14)
Caucasian	.08	–.16	.05	.53
	(.27)	(.29)	(.27)	(.27)
Income	–.16	–.07	.09	.13
	(.10)	(.10)	(.09)	(.10)
AIC	631	660	630	645
Observations	275	286	287	295

Note: Standard errors in parentheses; * $p <. 05$.

TABLE A6.7. Effect of Violating International Law Compared to May Violate International Law in Second Experiment

	Model 11 High Proportionality, High Distinction	Model 12 Low Proportionality, High Distinction	Model 13 High Proportionality, Low Distinction	Model 14 Low Proportionality, Low Distinction
Violates international law	.45	.35	1.10	.63
	(.23)	(.22)	(.24)	(.22)
Confidence in military	–.88*	–.61*	–.81*	–.67*
	(.16)	(.15)	(.17)	(.15)
Military experience	.35	.05	.27	–.44
	(.29)	(.30)	(.30)	(.30)
Partisanship	.31*	.21*	.24	.12
	(.13)	(.12)	(.13)	(.13)
Gender	.94*	.07	.24	.40
	(.26)	(.25)	(.26)	(.25)
Age	.02	–.01	.00	.00
	(.01)	(.01)	(.01)	(.01)
Education	.00	.33	–.01	–.12
	(.14)	(.14)	(.14)	(.13)
Caucasian	.05	–.21	–.05	.22
	(.25)	(.28)	(.27)	(.27)
Income	.01	.01	–.01	.13
	(.10)	(.09)	(.09)	(.09)
AIC	639	640	616	685
Observations	275	276	289	299

Note: Standard errors in parentheses; * $p <. 05$.

Notes

Chapter 1

1. Marcus Schulzke, "Ethically Insoluble Dilemmas in War," *Journal of Military Ethics* 12:2 (2013), 95–110; Marcus Schulzke, "The Unintended Consequences of War: Self-Defense and Violence against Civilians in Ground Combat Operations," *International Studies Perspectives* (forthcoming).

2. David Galula, *Counterinsurgency Warfare: Theory and Practice* (New York: Praeger, 1964); Frank Kitson, *Low Intensity Operations: Subversion, Insurgency and Peacekeeping* (London: Faber and Faber, 1971); Robert Thompson, *Defeating Communist Insurgency: The Lessons of Malaya and Vietnam* (New York: Praeger, 1966).

3. T. David Mason and Dale A. Krane, "The Political Economy of Death Squads: Toward a Theory of the Impact of State-Sanctioned Terror," *International Studies Quarterly* 33:2 (1989), 175–98; Sathis N. Kalyvas and Matthew A. Kocher, "How 'Free' Is Free Riding in Civil Wars?: Violence, Insurgency, and the Collective Action Problem," *World Politics* 59:2 (2007), 177–216; Sathis N. Kalyvas, *The Logic of Violence in Civil War* (New York: Cambridge University Press, 2006).

4. Matthew A. Kocher, Thomas B. Pepinsky, and Stathis N. Kalyvas, "Aerial Bombing and Counterinsurgency in the Vietnam War," *American Journal of Political Science* 55:2 (2011), 201–18; Luke N. Condra and Jacob N. Shapiro, "Who Takes the Blame? The Strategic Effects of Collateral Damage," *American Journal of Political Science* 56:1 (2010), 167–87; Jason Lyall, Graeme Blair, and Kosuke Imai, "Explaining Support for Combatants During Wartime: A Survey Experiment in Afghanistan," *American Political Science Review* 107:4 (2013), 679–705.

5. John Mueller, *War, Presidents, and Public Opinion* (New York: Wiley, 1973); Richard Eichenberg, "Victory Has Many Friends: U.S. Public Opinion and the Use of Force," *International Security* 30:1 (2005), 140–77; Matthew A. Baum and Tim Groeling, "Reality Asserts Itself: Public Opinion on Iraq and the Elasticity of Reality," *International Organization* 64 (2010), 443–79; C. V. Sirin, "Public Support for Military Interventions across Levels of Political Information and Stages of Intervention: The Case of the Iraq War," *Armed Forces & Society* 38:2 (August 2011), 252–72.

6. Michael C. Horowitz, Sarah E. Kreps, and Matthew Fuhrmann, "Separating Fact from Fiction in the Debate over Drone Proliferation," *International Security* 41:2 (2016), 7–42.

7. An important exception is Paul G. Gillespie, *The Development of Precision Guided Munitions* (Tuscaloosa: University of Alabama Press, 2006).

8. Robert Pape, *Bombing to Win: Air Power and Coercion in War* (Ithaca: Cornell University Press, 1996) reviews the historical debate about air power and war; for discussions of precision air power specifically, see David R. Lake, "The Limits of Coercive Airpower: NATO's 'Victory' in Kosovo Revisited," *International Security* 34:1 (2009), 83–112; Robert A. Pape, "The Limits of Precision-Guided Air Power," *Security Studies* 7:2 (1997), 93–114; Daryl G. Press, "The Myth of Air Power in the Persian Gulf War and the Future of Warfare," *International Security* 26:2 (2001), 5–44.

9. Charles Dunlap, "Technology: Recomplicating Moral Life for the Nation's Defenders," *Parameters* 29:3 (1999), 24–53; Harald Müller, "The Antinomy of the Democratic Peace," *International Politics* 41:4 (2004), 494–520; Jeffrey Record, "Collapsed Countries, Casualty Dread, and the New American Way of War," *Parameters* 32:2 (2002), 4–23.

10. James Corum and Wray R. Johnson, *Airpower in Small Wars: Fighting Insurgents and Terrorists* (Lawrence: University of Kansas Press, 2003).

11. David Omissi, *Air Power and Colonial Control: The Royal Air Force, 1919–1939* (Manchester: Manchester University Press, 1990); Robert Peck, "Aircraft in Small Wars," *Journal of the Royal United Services Academy* 73 (1928), 535–50; Mark Clodfelter, *The Limits of Air Power: The American Bombing of North Vietnam* (Omaha: University of Nebraska Press, 2006); Corum and Johnson, *Airpower in Small Wars*; Benjamin Lambeth, *The Transformation of American Air Power* (Ithaca: Cornell University Press, 2000).

12. Dennis Drew, *Insurgency and Counterinsurgency: American Military Dilemmas and Doctrinal Proposals* (Maxwell Air Force Base, AL: Air University Press, 1998); Thomas Thayer, *War Without Fronts* (Boulder: Westview Press, 1985); Raphael Littauer and Norman Uphoff, *The Air War in Indochina* (Boston: Beacon Press, 1972).

13. Mohammed M. Hafez and Joseph M. Hatfield, "Do Targeted Assassinations Work? A Multivariate Analysis of Israel's Controversial Tactic During Al-Aqsa Uprising," *Studies in Conflict & Terrorism* 29:4 (2006), 359–82; David Jaeger and M. Daniele Paserman, "The Cycle of Violence? An Empirical Analysis of Fatalities in the Palestinian-Israeli Conflict," *American Economic Review* 98:4 (2008), 1591–1604; Daniel Byman, "Do Targeted Killings Work?," *Foreign Affairs* (2002), 95–111; Steven R. David, "Fatal Choices: Israel's Policy of Targeted Killing," *Mideast Security and Policy Studies* 51:4 (2002).

14. Geoffry Blainey, *The Causes of War* (New York: Free Press, 1973); Bruce Bueno de Mesquita, *The War Trap* (New Haven: Yale University Press, 1975).

15. James D. Fearon, "Rationalist Explanations for War," *International Organization* 49:3 (1995), 379–414.

16. John Lewis Gaddis, *Strategies of Containment: A Critical Appraisal of American National Security Policy During the Cold War* (New York: Oxford University Press, 1995); Scott Sagan and Kenneth Waltz, *The Spread of Nuclear Weapons: An Enduring Debate*, 3rd. ed. (New York: W. W. Norton, 2013).

17. Constraints on drone operations still exist and limit the extent to which these weapons platforms can be used. See Andrea Gilli, Mauro Gilli, "The Diffusion of Drone Warfare? Industrial, Organizational, and Infrastructural Constraints," *Security Studies* 25:1 (2016), 50–84. These barriers are important to account for when estimating the likely extent of drone proliferation and predicting how drones will be used in future conflicts, but what is important for our purposes is that drones may be able to reduce the financial and material costs of war in ways that could lower inhibitions against fighting.

18. Micah Zenko and Sarah Kreps, *Limiting Armed Drone Proliferation* (New York: Council of Foreign Relations, 2014).

19. Important works include Gabriel Almond, *The American People and Foreign Policy* (New York: Praeger, 1950); Eichenberg, "Victory Has Many Friends," 140–77; Ole Holsti, *Public Opinion and American Foreign Policy*, rev. ed. (Ann Arbor: University of Michigan Press, 2007); Benjamin I. Page and Marshall M. Bouton, *The Foreign Policy Disconnect: What Americans Want from Our Leaders but Don't Get* (Chicago: University of Chicago Press, 2006); James N. Rosenau, *Public Opinion and Foreign Policy* (New York: Random House, 1961).

20. Bruce Jentleson, "The Pretty Prudent Public: Post Post-Vietnam American Opinion on the Use of Military Force," *International Studies Quarterly* 36:1 (1992), 49–74.

21. Matthew A. Baum and Timothy Groeling, "Reality Asserts Itself: Public Opinion on Iraq and the Elasticity of Reality," *International Organization* 64:3 (2010), 443–79; William A. Boettcher and Michael D. Cobb, "Echoes of Vietnam?: Casualty Framing and Public Perceptions of Success and Failure in Iraq," *Journal of Conflict Resolution* 50:6 (2006), 831–54; Scott S. Gartner and Gary M. Segura, "War, Casualties, and Public Opinion," *Journal of Conflict Resolution* 42:3 (1998), 278–300; Douglas Kriner and Francis X. Shen, "How Citizens Respond to Combat Casualties: The Differential Impact of Local Casualties on Support for the War in Afghanistan," *Public Opinion Quarterly* 76:4 (2012), 761–70; John Mueller, *War, Presidents, and Public Opinion* (New York: Wiley, 1973).

22. See especially Christopher Gelpi, Peter Feaver, and Jason Reifler, *Paying the Human Costs of War: American Public Opinion and Casualties in Military Conflicts* (Princeton: Princeton University Press, 2009); and Eric V. Larson, *Casualties and Consensus: The Historical Role of Casualties in Domestic Support for Wars and Military Operations* (Santa Monica, CA: RAND, 1996).

23. Robert Johns and Graeme A. M. Davies, "Civilian Casualties and Public Support for Military Action: Experimental Evidence," *Journal of Conflict Resolution* (forthcoming), John Mueller, "Public Opinion as a Constraint on U.S. Foreign Policy: Assessing the Perceived Value of American and Foreign Lives" (manuscript, 2000); John Tirman, *The Deaths of Others: The Fate of Civilians in America's Wars* (New York: Oxford University Press, 2011); Eichenberg, "Victory Has Many Friends."

24. John Zaller, *The Nature and Origin of Mass Opinion* (New York: Cambridge University Press, 1992); and Adam Berinsky, "Assuming the Costs of War: Events, Elites, and American Public Support for Military Conflict," *Journal of Politics* 69:4 (2007), 975–97; on the role of the media, see Shanto Iyengar, "Television News and Citizens' Explanations of National Affairs," *American Political Science Review* 81:3

(1987), 815–32; and Hector Pera, "Explaining Public Support for the Use of Military Force: The Impact of Reference Point Framing and Prospective Decision Making," *International Organization* 65:1 (2011), 139–67; on international actors, see Danny Hayes and Matt Guardino, *Influence from Abroad: Foreign Voices, the Media, and U.S. Public Opinion* (New York: Cambridge University Press, 2013). A useful discussion of this approach is Joshua D. Kertzer and Thomas Zeitzoff, "A Bottom-Up Theory of Public Opinion About Foreign Policy," *American Journal of Political Science* 61:3 (2017), 543–58.

25. Jon Hurwitz and Mark A. Peffley, "How Are Foreign Policy Attitudes Structured? A Hierarchical Model," *American Political Science Review* 81 (1987), 1099–1120; Joshua D. Kertzer, "Making Sense of Isolationism: Foreign Policy Mood as a Multilevel Phenomenon," *Journal of Politics* 75 (2013), 225–40; Brian C. Rathbun, "Hierarchy and Community and Foreign Policy Beliefs in American Elites," *Journal of Conflict Resolution* 51 (2007), 379–407; Eugene R. Wittkopf, "The Structure of Foreign Policy Attitudes: An Alternative View," *Social Science Quarterly* 62 (1981), 108–23.

26. Paul Fahlstrom and Thomas Gleason, *Introduction to UAV Systems* (Hoboken, NJ: John Wiley, 2012).

27. Fahlstrom and Gleason, *Introduction to UAV Systems*, 5.

28. Fahlstrom and Gleason, *Introduction to UAV Systems*, 5.

29. Thomas P. Ehrhard, *Air Force UAVs: The Secret History* (Arlington, VA: Mitchell Institute for Airpower Studies, 2010).

30. Elizabeth Becker, "Crisis in the Balkans: The Drones," *New York Times*, June 3, 1999, http://www.nytimes.com/1999/06/03/world/crisis-balkans-drones-they-re-unmanned-they-fly-low-they-get-picture.html

31. Chris Woods, "The Story of America's First Drone Strike," *The Atlantic*, May 30, 2015, http://www.theatlantic.com/international/archive/2015/05/america-first-drone-strike-afghanistan/394463/

32. Chris Woods, *Sudden Justice: The True Costs of America's Secret Drone War* (London: C. Hurst & Company, 2014), 25.

33. Woods, "The Story of America's First Drone Strike."

34. P. W. Singer, *Wired for War: The Robotics Revolution and Conflict in the 21st Century* (New York: Penguin Press, 2009), 35.

35. Singer, *Wired for War*, 34.

36. https://www.thebureauinvestigates.com/2015/02/02/almost-2500-killed-covert-us-drone-strikes-obama-inauguration/

37. Woods, *Sudden Justice*, 55–61.

38. http://www.longwarjournal.org/multimedia/Yemen/code/Yemen-strike.php

39. https://www.thebureauinvestigates.com/2015/10/05/monthly-drone-report-total-drone-strikes-under-obama-in-pakistan-somalia-and-yemen-now-491-after-september-attacks/

40. https://www.thebureauinvestigates.com/2015/10/05/monthly-drone-report-total-drone-strikes-under-obama-in-pakistan-somalia-and-yemen-now-491-after-september-attacks/

41. Eric Schmitt and David E. Sanger, "Pakistan Shift Could Curtail Drone

Strikes," *New York Times*, February 22, 2008, http://www.nytimes.com/2008/02/22/washington/22policy.html

42. Greg Miller, "CIA Seeks to Expand Drone Fleet, Officials Say," *Washington Post*, April 18, 2012, http://www.washingtonpost.com/world/national-security/cia-seeks-new-authority-to-expand-yemen-drone-campaign/2012/04/18/gIQAsaumRT_print.html

43. Scott Shane and Jo Becker, "Secret 'Kill List' Tests Obama's Principles," *New York Times*, May 29, 2012, http://www.nytimes.com/2012/05/29/world/obamas-leadership-in-war-on-al-qaeda.html?_r=1

44. Jonathan Landay, "Obama's Drone War Kills 'Others,' Not Just Al Qaeda Leaders," *McClatchy*, April 9, 2013.

45. Daniel Klaidman, "Drones: How Obama Learned to Kill," *Newsweek*, May 28, 2012, http://www.thedailybeast.com/newsweek/2012/05/27/drones-the-silent-killers.html

46. Mark Mazzetti, "The Drone Zone," *New York Times Magazine*, July 6, 2012: http://www.nytimes.com/2012/07/08/magazine/the-drone-zone.html?pagewanted=all

47. Akbar Ahmed and Frankie Martin, "Deadly Drone Strike on Muslims in the Southern Philippines," *Brookings*, March 5, 2012, http://www.brookings.edu/research/opinions/2012/03/05-drones-philippines-ahmed

48. https://www.wired.com/2012/11/drones-afghan-air-war/

49. Alice Ross, "Who Is Dying in Afghanistan's 1,000-plus Drone Strikes?," *Bureau of Investigative Journalism*, July 2014: https://www.thebureauinvestigates.com/2014/07/24/who-is-dying-in-afghanistans-1000-plus-drone-strikes/

50. http://www.publications.parliament.uk/pa/cm201213/cmhansrd/cm121106/halltext/121106h0001.htm#12110684000252

51. http://securitydata.newamerica.net/drones/somalia-analysis.html

52. https://www.thebureauinvestigates.com/2012/02/22/get-the-data-somalias-hidden-war/

53. Richard Osley, "Abu Nabil: Isis Leader in Libya 'killed by US drone strike,'" *The Independent*, November 14, 2015, http://www.independent.co.uk/news/world/africa/abu-nabil-isis-leader-in-libya-killed-by-us-drone-strike-a6734961.html

54. Adam Grissom, "The Future of Military Innovation Studies," *Journal of Strategic Studies* 29:5 (2006), 905–34.

55. Michael C. Horowitz, Sarah E. Kreps, Matthew Fuhrmann, "Separating Fact from Fiction in the Debate over Drone Proliferation," *International Security* 41:2 (2016), 7–42.

56. Andrew Hill, "Military Innovation and Military Culture," *Parameters* 45:1 (2015), 85–98.

57. Stephen P. Rosen, *Winning the Next War: Innovation and the Modern Military* (Ithaca: Cornell University Press, 1991).

58. For more about the influence of institutional culture on innovations, see Elizabeth Kier, *Imagining War: French and British Military Doctrine Between the Wars* (Princeton: Princeton University Press, 1997).

59. Barry R. Posen, *The Sources of Military Doctrine: France, Britain, and Germany Between the World Wars* (Ithaca: Cornell University Press, 1986), 56.

60. Peter R. Mansoor, *Surge: My Journey with General David Petraeus and the Remaking of the Iraq War* (New Haven: Yale University Press, 2013).

61. Matt J. Martin and Charles W. Sasser, *Predator: The Remote-Control Air War Over Iraq and Afghanistan: A Pilot's Study* (Minneapolis, MN: Zenith Press, 2010), 205–14.

62. John D. Sutter, "How 9/11 Inspired a New Era of Robotics," *CNN*, September 7, 2011: http://edition.cnn.com/2011/TECH/innovation/09/07/911.robots.disaster.response/

63. http://www.af.mil/AboutUs/FactSheets/Display/tabid/224/Article/104516/rq-4-global-hawk.aspx

64. MQ-9 Reapers Add to Arsenal with First GBU-38 Drop. U.S. Air Force, May 8, 2017, http://www.af.mil/News/Article-Display/Article/1175849/mq-9-reapers-add-to-arsenal-with-first-gbu-38-drop/

65. Sarah Kreps, "Flying under the Radar: A Study of Public Attitudes Towards Unmanned Aerial Vehicles," *Research and Politics* 1:1 (2014).

66. Paul Sniderman and Thomas Piazza, *The Scar of Race* (Cambridge, MA: Harvard University Press, 1993); Paul Sniderman and Douglas B. Grob, "Innovations in Experimental Design in Attitude Surveys," *Annual Review of Sociology* 22 (1996), 377–99; Brian J. Gaines and James H. Kuklinski, "The Logic of the Survey Experiment Re-Examined," *Political Analysis* 15:1 (2007), 1–15.

67. Zaller, *Nature and Origins of Mass Opinion*; John Zaller and Stanley Feldman, "A Simple Theory of the Survey Response: Answering Questions versus Revealing Preferences," *American Journal of Political Science* 36:3 (1992), 579–616.

68. Neil Malhotra and Jon Krosnick, "The Effect of Survey Mode and Sampling on Inferences about Political Attitudes and Behavior: Comparing the 2000 and 2004 ANES to Internet Surveys with Nonprobability Samples," *Political Analysis* 15:3 (2007), 286–323; and Stephen Ansolabehere and Brian F. Schaffner, "Does Survey Mode Still Matter? Findings from a 2010 Multi-Mode Comparison," *Political Analysis* 22:3 (2014), 285–303.

69. Adam J. Berinsky, Gregory A. Huber, and Gabriel B. Lenz, "Evaluating Online Labor Markets for Experimental Research: Amazon.com's Mechanical Turk," *Political Analysis* 20:3 (2012), 351–68.

70. Kevin J. Mullinix, Thomas J. Leeper, James N. Druckman, and Jeremy Freese, "The Generalizability of Survey Experiments," *Journal of Experimental Political Science* 2:2 (2015), 109–38.

71. Connor Huff and Dustin Tingley, "'Who Are These People?' Evaluating the Demographic Characteristics and Political Preferences of MTurk Survey Respondents," *Research & Politics* 4:1 (2015), 1–7; Kevin E. Levay, Jeremy Freese, and James N. Druckman, "The Demographic and Political Composition of Mechanical Turk Samples," *Sage Open* 6:1 (2016).

72. Data files and scripts used to clean and analyze this data are available at http://www.jamesigoewalsh.com

73. We preregistered these designs with the registry AsPredicted.org; see chapters 2 and 3 for details.

74. For discussions of preregistration, see Katherine Casey, Racheel Glennerster, and Edward Miguel, *Reshaping Institutions: Evidence on Aid Impacts Using a Pre-Analysis*

Plan (National Bureau of Economic Research working paper no. 17012, 2011); Macartan Humphreys, Raul Sanchez de la Sierra, and Peter van der Windt, "Fishing, Commitment, and Communication: A Proposal for Comprehensive Nonbinding Research Registration," *Political Analysis* 21:1 (2013), 1–20.

Chapter 2

1. John E. Mueller, "The Search for the 'Breaking Point' in Vietnam: The Statistics of a Deadly Quarrel." *International Studies Quarterly* 23 (1979), 301–29; John E. Mueller, "Public Opinion as a Constraint on U.S. Foreign Policy: Assessing the Perceived Value of American and Foreign Lives," paper prepared for the annual meeting of the International Studies Association, Los Angeles, March 2000; John E. Mueller, "The Iraq Syndrome," *Foreign Affairs* 86:4 (2005), 44–54; Edward N. Luttwak, "Toward Post-Heroic Warfare," *Foreign Affairs* (May/June 2005).

2. Michael Ignatieff, *Virtual War: Kosovo and Beyond* (New York: Picador, 2001); Christopher Coker, *Waging War Without Warriors? The Changing Culture of Military Conflict* (Boulder: Lynne Rienner, 2000).

3. Eric V. Larson, *Casualties and Consensus: The Historical Role of Casualties in Domestic Support for U.S. Military Operations* (Santa Monica, CA: RAND, 1996), 101; Charles K. Hyde, *Casualty Aversion: Implications for Policy Makers and Senior Military Officers* (Newport, RI: Naval War College, 2000); Christopher Gelpi, Jason Reifler, Peter Feaver, "Iraq the Vote: Retrospective and Prospective Foreign Policy Judgments on Candidate Choice and Casualty Tolerance," *Political Behavior* 29 (2007), 151–74.

4. Daniel Byman, "Why Drones Work: The Case for Washington's Weapon of Choice," *Foreign Affairs* (2013), 1–9.

5. Colin H. Kahl, "In the Crossfire or the Crosshairs? Norms, Civilian Casualties, and U.S. Conduct in Iraq," *International Security* 32:1 (2007), 7–46.

6. Declan Walsh, "Drone Strikes on Al Qaeda Are Said to Take Toll on Leadership in Pakistan," *New York Times*, April 24, 2015; Spencer Ackerman, "Obama Claims US Drones Strikes Have Killed Up to 116 Civilians," *The Guardian*, July 1, 2016.

7. A. C. Grayling, *Among the Dead Cities: The History and Moral Legacy of the WWII Bombing of Civilians in Germany and Japan* (New York: Walker & Company, 2006).

8. Peter W. Singer, *Wired for War: The Robotics Revolution and Conflict in the 21st Century* (New York: Penguin Press, 2009).

9. John Kaag and Sarah Kreps, *Drone Warfare* (Malden, MA: Polity Press, 2014).

10. Maja Zehfuss, "Targeting: Precision and the Production of Ethics," *European Journal of International Relations* 17:3 (2011), 543–66.

11. James Der Derian, *Virtuous War: Mapping the Military-Industrial-Media-Entertainment Network* (New York: Routledge, 2009); John Tirman, *The Deaths of Others: The Fate of Civilians in America's Wars* (New York: Oxford University Press, 2011).

12. Sarah Kreps, "Flying under the Radar: A Study of Public Attitudes Towards Unmanned Aerial Vehicles," *Research and Politics* 1:1 (2014).

13. Larson, *Casualties and Consensus.*

14. See especially Christopher Gelpi, Peter Feaver, and Jason Reifler, *Paying the Human Costs of War: American Public Opinion and Casualties in Military Conflicts* (Princeton: Princeton University Press, 2009); see also Steven Kull and Clay Ramsay, "The Myth of the Reactive Public: American Public Attitudes on Military Fatalities in the Post-Cold War Period," in *Public Opinion and The International Use of Force*, ed. Phillip Everts and Pierangelo Isneria (London: Routledge, 2001), 205–28.

15. Adam J. Berinsky, "Assuming the Costs of War: Events, Elites, and American Public Support for Military Conflict," *Journal of Politics* 69 (2006), 975–97; and Adam J. Berinsky and James N. Druckman, "The Polls—Review Public Opinion Research and Support for the Iraq War," *Public Opinion Quarterly* 71 (2007), 126–41.

16. Matthew A. Baum and Tim Groeling, "Reality Asserts Itself: Public Opinion on Iraq and the Elasticity of Reality," *International Organization* 64 (2010), 443–79.

17. Mueller, "Public Opinion as a Constraint on U.S. Foreign Policy"; and Tirman, *The Deaths of Others.*

18. Robert Johns and Graeme A. M. Davies, "Civilian Casualties and Public Support for Military Action: Experimental Evidence," *Journal of Conflict Resolution* (forthcoming); and Benjamin Valentino, "Moral Character or Character of War? American Public Opinion on the Targeting of Civilians in Time of War," *Daedalus* 145:4 (2016), 127–38.

19. Richard Eichenberg, "Victory Has Many Friends: U. S. Public Opinion and the Use of Force," *International Security* 30:1 (2005), 140–77.

20. Gelpi, Feaver, and Reifler, *Paying the Human Costs of War.*

21. Bruce Jentleson, "The Pretty Prudent Public: Post Post-Vietnam American Opinion on the Use of Military Force," *International Studies Quarterly* 36 (1992), 49–73.

22. Bruce W. Jentleson and Rebecca L. Britton, "Still Pretty Prudent: Post-Cold War American Public Opinion on the Use of Military Force," *Journal of Conflict Resolution* 42:4 (1998), 395–417.

23. Gelpi, Feaver, and Reifler, *Paying the Human Costs of War.*

24. A. Cooper Drury, L. Marvin Overbury, Adrian Ang, Yitan Li, "'Pretty Prudent' or Rhetorically Responsive? The American Public's Support for Military Action," *Political Research Quarterly* 63:1 (2010), 92.

25. The seminal work is John Zaller, *The Nature and Origins of Mass Opinion* (New York: Cambridge University Press, 1992).

26. For this process, see Arthur Lupia and Matthew McCubbins, *The Democratic Dilemma: Can Citizens Learn What They Need to Know* (New York: Cambridge University Press, 1998); and Samuel L. Popkin, *The Reasoning Voter: Communication and Persuasion in Presidential Campaigns* (Chicago: University of Chicago Press, 1994); for applications to foreign policy, see Alexander Thompson, "Coercion through IOs: The Security Council and the Logic of Information Transmission," *International Organization* 60 (2006), 1–34; and Danny Hayes and Matt Guardino, *Influence from Abroad: Foreign Voices, the Media, and U.S. Public Opinion* (New York: Cambridge University Press, 2013); for discussion in the context of drone strikes, see Kreps, "Flying under the Radar."

27. Jens Hainmueller, Daniel J. Hopkins, and Teppei Yamamoto, "Causal Inference in Conjoint Analysis: Understanding Multidimensional Choices via Stated Preference

Experiments," *Political Analysis* 22:1 (2014), 1–30; Jens Hainmueller and Daniel J. Hopkins, "The Hidden American Immigration Consensus: A Conjoint Analysis of Attitudes Toward Immigrants," *American Journal of Political Science* 59 (2015), 529–48; and Jens Hainmueller, Dominik Hangartner, and Teppei Yammamoto, "Validating Vignette and Conjoint Survey Experiments Against Real-World Behavior," *Proceedings of the National Academy of Sciences* 112:8 (2015), 2395–2400.

28. We preregistered this experiment; the pre-analysis plan is available at https://aspredicted.org/6ivav.pdf

29. The experiment was fielded April 25–30, 2016. See table A2.2 in the chapter appendix for a comparison of demographic variables from this sample with those from the American National Election Studies.

30. See Hainmueller, Hopkins, and Yamamoto, "Causal Inference in Conjoint Analysis."

31. Hainmueller, Hopkins, and Yamamoto, "Causal Inference in Conjoint Analysis."

32. For another study that reaches similar conclusions on this score, see Jacquelyn Schneider and Julia Macdonald, *U.S. Public Support for Drone Strikes: When Do Americans Prefer Unmanned over Manned Platforms?* (Washington, DC: Center for a New American Security, 2015).

33. Gelpi, Feaver, and Reifler, *Paying the Human Costs of War*, 67–124; Ole Holsti, *Public Opinion and American Foreign Policy*, rev. ed. (Ann Arbor: University of Michigan Press, 2004); for a discussion of political ideology and foreign policy, see Peter Hays Gries, *The Politics of American Foreign Policy: How Ideology Divides Liberals and Conservatives over Foreign Affairs* (Stanford, CA: Stanford University Press, 2014).

34. This was also the case when we examined ratings of each attack type individually. Democrats provided lower ratings than did Republicans for drone strikes, air strikes, and the use of ground troops. See the replication materials for details.

35. Discussions of drone policy include Hugh Gusterson, *Drone: Remote Control Warfare* (Cambridge: MIT Press, 2016); Sarah Kreps, *Drones: What Everyone Needs to Know* (New York: Oxford University Press, 2016); and Scott Shane, *Objective Troy: A Terrorist, A President, and the Rise of the Drone* (New York: Random House, 2015); see also James Igoe Walsh, "The Rise of Targeted Killings," *Journal of Strategic Studies*.

36. Kreps, "Flying under the Radar."

37. Christopher Gelpi, Peter D. Feaver, and Jason Reifler, "Success Matters: Casualty Sensitivity and the War in Iraq." *International Security* 30:3 (2006), 7–46; Gelpi, Feaver, and Reifler, *Paying the Human Costs of War*.

38. John Mueller, *Overblown: How Politicians and the Terrorism Industry Inflate National Security Threats, and Why We Believe Them* (New York: Simon & Schuster, 2009); John Mueller and Mark G. Stewart, "The Terrorism Delusion: America's Overwrought Response to September 11," *International Security* 37:1 (2012), 81–110.

39. Michael C. Horowitz, Sarah E. Kreps, and Matthew Fuhrmann, "Separating Fact from Fiction in the Debate over Drone Proliferation," *International Security* 41:2 (2016).

40. Helen Milner and Dustin Tingley "The Domestic Politics of Foreign Aid: American Legislators and the Politics of Donor Countries," *Economics and Politics* 22:2 (2010), 200–232; Helen Milner and Dustin Tingley, "Who Supports Global

Economic Engagement? The Sources of Preferences in American Foreign Economic Policy," *International Organization* 65:1 (2011), 37–68.

41. Joshua W. Busby and Jonathan Monten, "Without Heirs? Assessing the Decline of Establishment Internationalism in U.S. Foreign Policy," *Perspectives on Politics* 6:3 (2008), 451–72.

42. Table A2.1 in the appendix presents the results of the model depicted in figure 2.2. As a robustness check, it also reports a model using the rating variable. This has been transformed so that values of 4 or less on the seven-point rating scale are equal to zero, and values of 5 or greater are equal to 1. (This approach is suggested in Hainmueller and Hopkins, "The Hidden American Immigration Consensus.") The findings from using this dependent variable are largely consistent with those in which the preference dependent variable is employed. The one difference to note is the results for international support, where the Security Council opposes and the human rights NGOs support and oppose variables are no longer statistically significant.

43. Frank Newport, "Seventy-Two Percent of Americans Support War Against Iraq," *Gallup*, March 24, 2003, http://www.gallup.com/poll/8038/seventytwo-percent-americans-support-war-against-iraq.aspx; Pew Research Center, "Public Attitudes Toward the War in Iraq: 2003–2008," March 19, 2008, http://www.pewresearch.org/2008/03/19/public-attitudes-toward-the-war-in-iraq-20032008/

44. Will Dahlgreen, "Memories of Iraq: Did We Ever Support the War?," *YouGov UK*, June 3, 2013, https://yougov.co.uk/news/2015/06/03/remembering-iraq/

45. $\chi^2 = 68.12$, degrees of freedom = 4, $p < .01$.

46. Adam J. Berinsky, *In Time of War: Understanding American Public Opinion from World War II to Iraq* (Chicago: University of Chicago Press, 2009).

47. Aaron M. Hoffman, Christopher R. Agnew, Laura E. VanderDrift, and Robert Kulzick, "Norms, Diplomatic Alternatives, and the Social Psychology of War Support," *Journal of Conflict Resolution* 59:1 (2013), 3–28; Benjamin I. Page and Marshal M. Bouton, *The Foreign Policy Disconnect: What Americans Want from Our Leaders But Don't Get* (Chicago: University of Chicago Press, 2006).

48. Harvey Sapolsky and Jeremy Shapiro, "Casualties, Technology, and America's Future Wars," *Parameters* (1996), 119–27.

49. Kaag and Kreps, *Drone Warfare*; Christian Enemark, *Armed Drones and the Ethics of War: Military Virtue in a Post-Heroic Age* (New York: Routledge, 2013); Grégorie Chamayou, *A Theory of the Drone*, trans. Janet Lloyd (New York: The New Press, 2015).

Chapter 3

1. Matthew A. Baum and Tim Groeling, "Reality Asserts Itself: Public Opinion on Iraq and the Elasticity of Reality," *International Organization* 64 (2010), 443–79; John Mueller, *War, Presidents, and Public Opinion* (New York: Wiley, 1973); Scott Sigmund Gartner, "The Multiple Effects of Casualties on Public Support for War: An Experimental Approach," *American Political Science Review* 102:1 (2008), 95–106.

2. Christopher Gelpi, Peter Feaver, and Jason Reifler, *Paying the Human Costs of War: American Public Opinion and Casualties in Military Conflicts* (Princeton: Princeton University Press, 2009).

3. Gartner, "The Multiple Effects of Casualties on Public Support for War."

4. Peter W. Singer, "Do Drones Undermine Democracy?" *New York Times*, January 21, 2012.

5. Gelpi, Reifler, and Feaver (*Paying the Human Costs of War*) distinguish between two different manifestations of this feeling. "Casualty sensitivity recognizes the human toll as a cost of war; casualty phobia refers to a sensitivity so great that it amounts to an unwillingness to support a military operation if even very low human costs are incurred." What we refer to as "casualty aversion" is an opposition to a prospective or ongoing war that is informed by concern over military casualties, regardless of the extent of the anticipated casualties. Thus, unlike Gelpi, Reifler, and Feaver, we want to avoid suggesting that this phenomenon is linked to low casualty levels and to remain open to opposition being triggered by variations in the number of casualties. We also prefer to describe this as "casualty aversion" to avoid any pejorative implications of the term "phobia."

6. John E. Mueller, "Trends in Popular Support for the Wars in Korea and Vietnam," *American Political Science Review* 65 (June 1971), 358–375; John Mueller, *War, Presidents, and Public Opinion* (New York: Wiley, 1973); John E. Mueller, "The Search for the 'Breaking Point' in Vietnam: The Statistics of a Deadly Quarrel," *International Studies Quarterly* 23 (1979), 301–29.

7. Jeffrey S. Milstein, *Dynamics of the Vietnam War: A Quantitative Analysis and Predictive Computer Simulation* (Columbus: Ohio State University Press, 1974).

8. William C. Martel, *Victory in War: Foundations of Modern Strategy* (Cambridge: Cambridge University Press, 2011), 163.

9. John E. Mueller, "Policy Principles for Unthreatened Wealth-Seekers," *Foreign Policy* 102 (Spring 1996), 31.

10. Jacob Weisberg, "Zero Tolerance," *New York* 27:40 (October 10, 1994), 21.

11. Harvey Sapolsky and Jeremy Shapiro, "Casualties, Technology, and America's Future Wars," *Parameters* (1996), 119–27.

12. John E. Mueller, "The Iraq Syndrome," *Foreign Affairs* 86:4 (2005), 44–54, 44.

13. Edward N. Luttwak, "Toward Post-Heroic Warfare," *Foreign Affairs,* (May/June 1995).

14. James Burk, "Public Support for Peacekeeping in Lebanon and Somalia: Assessing the Casualties Hypothesis," *Political Science Quarterly* 114:1 (1999), 53–78, 77.

15. Eric V. Larson, *Casualties and Consensus: The Historical Role of Casualties in Domestic Support for U.S. Military Operations* (Santa Monica, CA: RAND, 1996), 101.

16. Charles K. Hyde, *Casualty Aversion: Implications for Policy Makers and Senior Military Officers* (Newport, RI: Naval War College, 2000).

17. Christopher Gelpi, Jason Reifler, and Peter Feaver, "Iraq the Vote: Retrospective and Prospective Foreign Policy Judgments on Candidate Choice and Casualty Tolerance," *Political Behavior* 29 (2007), 151–74.

18. D. L. Kriner and F. X. Shen, "Reassessing American Casualty Sensitivity: The Mediating Influence of Inequality," *Journal of Conflict Resolution* 58 (2013), 1174–1201.

19. Scott Sigmund Gartner, Gary M. Segura, and Michael Wilkening, "All Politics Are Local: Local Losses and Individual Attitudes toward the Vietnam War," *Journal of Conflict Resolution* 41:5 (1997), 669–94.

20. Douglas Delaney, "Cutting, Running, or Otherwise? The US Decision to Withdraw from Somalia," *Small Wars & Insurgencies* 15:3 (2004), 28–48.

21. William A. Boettcher III and Michael D. Cobb, "'Don't Let Them Die in Vain': Casualty Frames and Public Tolerance for Escalating Commitment in Iraq," *Journal of Conflict Resolution* 53:5 (2009), 677–97.

22. Peter R. Mansoor, *Surge: My Journey with General David Petraeus and the Remaking of the Iraq War* (New Haven: Yale University Press, 2013).

23. Alan P. Dobson, Steve Marsh, *US Foreign Policy since 1945* (New York: Routledge, 2001), 121–23; Richard Melanson, *American Foreign Policy Since the Vietnam War: The Search for Consensus from Richard Nixon to George W. Bush* (London: M.E. Sharpe, 2005), 287. See also Holsti's discussion of casualty sensitivity and how it is blamed for influencing foreign policy in Ole Holsti, *Making American Foreign Policy* (New York: Routledge, 2006).

24. Hugh Smith, "What Costs Will Democracies Bear? A Review of Popular Theories of Casualty Aversion," *Armed Forces & Society* 31:4 (2005), 487–512, 489.

25. Bradley Jay Strawser, "Moral Predators: The Duty to Employ Uninhabited Aerial Vehicles," *Journal of Military Ethics* 9:4 (2010), 342–68.

26. Sapolsky and Shapiro, "Casualties, Technology, and America's Future Wars."

27. Gelpi, Reifler, and Feaver, "Iraq the Vote," 156.

28. Charles K. Hyde, *Casualty Aversion: Implications for Policy Makers and Senior Military Officers* (Newport, RI: Naval War College, 2000).

29. Hyde, *Casualty Aversion*, 19.

30. Douglas Kriner and Francis Shen, "Responding to War on Capitol Hill: Battlefield Casualties, Congressional Response, and Public Support for the War in Iraq," *American Journal of Political Science* 58:1 (2014), 157–74.

31. Steven Kull and I. M. Destler, *Misreading the Public: The Myth of a New Isolationism* (Washington, DC: Brookings Institution, 1999).

32. John A. Gentry, *How Wars are Won and Lost: Vulnerability and Military Power* (Santa Barbara, CA: Praeger, 2012), 118.

33. Charles Krauthammer, "America, Battle-Tested," *Washington Post*, January 18, 2002, https://www.washingtonpost.com/archive/opinions/2002/01/18/america-battle-tested/9fc9e535-28e6-400b-b220-bcf050ce6e55/

34. Gentry, *How Wars Are Won and Lost*, 118.

35. Sapolsky and Shapiro, "Casualties, Technology, and America's Future Wars."

36. Hyde, *Casualty Aversion*; Richard A. Lacquement, *The Casualty-Aversion Myth* (Newport, RI: Naval War College, 1997).

37. P. W. Singer, *Wired for War: The Robotics Revolution and Conflict in the 21st Century* (New York: Penguin Press, 2009), 255.

38. John Kaag and Sarah Kreps, *Drone Warfare* (Malden, MA: Polity Press, 2014), 76.

39. For additional examples of this argument, see Grégorie Chamayou, *A Theory of the Drone* (New York: The New Press, 2015); David Hastings Dunn, "Drones: Disembodied Aerial Warfare and the Unarticulated Threat," *International Affairs* 89:5 (2013), 1237–46; Christian Enemark, *Armed Drones and the Ethics of War: Military Virtue in a Post-Heroic Age* (New York: Routledge, 2013).

40. Gelpi, Feaver, and Reifler, *Paying the Human Costs of War*.

41. For examples see https://www.rescue.org/article/ending-famine-defining-global-issue-millennials; Shibley Telhami, "American Attitudes toward the Middle East," *The Center for Middle East Policy at the Brookings Institution*, July 11, 2016. https://www.brookings.edu/wp-content/uploads/2016/07/FINAL-KEY-FINDINGS-3.pdf

42. Rose McDermott, *Risk-Taking in International Politics: Prospect Theory in American Foreign Policy* (Ann Arbor: University of Michigan Press, 1998).

43. Héctor Perla Jr., "Explaining Public Support for the Use of Military Force," *International Organization* 65:1 (2011), 139–67.

44. Sarah Kreps, "Flying under the Radar: A Study of Public Attitudes Towards Unmanned Aerial Vehicles," *Research and Politics* 1:1 (2014); Michael C. Horowitz, "Public Opinion and the Politics of the Killer Robot Debate," *Research & Politics* 3:1 (2014).

45. Perla, "Explaining Public Support for the Use of Military Force."

46. See the chapter appendix for the full text of the treatment conditions, the survey items, and demographic profiles of the two respondent pools.

47. The YouGov survey was fielded April 8–13, 2016; the Mechanical Turk survey was fielded May 25, 2016. During the period between when these experiments were fielded, there were no unusual or prominent developments concerning the use of armed drones that received significant news coverage.

48. Respondents in the YouGov sample were matched to a sampling frame on gender, age, race, education, party identification, ideology, and political interest. The frame was constructed by stratified sampling from the full 2010 American Community Survey (ACS) sample with selection within strata by weighted sampling with replacements (using the person weights on the public use file). Data on voter registration status and turnout were matched to this frame using the November 2010 Current Population Survey. Data on interest in politics and party identification were then matched to this frame from the 2007 Pew Religious Life Survey. The matched cases were weighted to the sampling frame using propensity scores. The matched cases and the frame were combined and a logistic regression was estimated for inclusion in the frame. The propensity score function included age, gender, race/ethnicity, years of education, and ideology. The propensity scores were grouped into deciles of the estimated propensity score in the frame and poststratified according to these deciles.

49. Preanalysis plans were filed with the registry aspredicted.org; the plan for the YouGov version of the experiment is available at https://aspredicted.org/6tb4m.pdf, and the plan for the Mechanical Turk version is available at https://aspredicted.org/rxj49.pdf

50. Kevin E. Levay, Jeremy Freese, and James N. Druckman, "The Demographic and Political Composition of Mechanical Turk Samples," *SAGE Open* 6 (2016); Kevin J. Mullinix, Thomas J. Leeper, James N. Druckman, and Jeremy Freese, "The Generalizability of Survey Experiments," *Journal of Experimental Political Science* 2 (2015), 109–38.

51. Adam J. Berinsky, Gregory A. Huber, and Gabriel S. Lenz, "Evaluating Online Labor Markets for Experimental Research: Amazon. Com's Mechanical Turk," *Political Analysis* 20 (2012), 351–68.

52. Levay, Freese, and Druckman, "The Demographic and Political Composition of Mechanical Turk Samples."

53. $\chi^2 = 1.92$, degrees of freedom = 4, $p = .75$. We collapsed the seven-point measure of support for the attack to three levels for this figure.

54. Gelpi, Feaver, and Reifler, *Paying the Human Costs of War.*

55. For more on the context of terrorist attacks during the period we conducted the survey experiments, see David Kilcullen, *Blood Year: The Unraveling of Western Counterterrorism* (Oxford: Oxford University Press, 2016).

56. $\chi^2 = 44.45$, degrees of freedom = 4, $p < .01$.

57. $\chi^2 = .59$, degrees of freedom = 4, $p = .96$.

58. $\chi^2 = .48$, degrees of freedom = 4, $p = .98$.

59. Some respondents assigned to the drone strike condition indicated that they expected military casualties, despite being informed that military personnel would not be at risk. There are a number of explanations for this. Some respondents might have thought that the remote pilots of the drone aircraft could suffer psychological harm, a possibility that has been frequently discussed in the media. Others might have thought about the longer-term consequences of the use of military force, worrying that seemingly low-risk attacks on the terrorist bases might lead to the introduction of more forces in the future, including ground troops who could be at greater risk of harm. Other research finds that there is a subset of individuals who strongly oppose the use of force under any conditions and that this preference negatively colors their assessments of the likely consequences of military action. On this point, see Feaver, Gelpi, and Reifler, *Paying the Human Costs of War.* Another explanation is that these respondents did not pay attention to the details of the treatment vignette or the question. It is possible that lower levels of attention would alter the relationships reported in table 3.2. To assess this possibility, we reran model 2 in table 3.2, but excluded respondents assigned to the drone strike treatment who indicated that military casualties were likely to result. Results are similar to those in model 2. See the replication materials for details.

60. John E. Mueller, "The Iraq Syndrome," *Foreign Affairs* 86:4 (2005), 53.

61. James Burk, "Public Support for Peacekeeping in Lebanon and Somalia: Assessing the Casualties Hypothesis," *Political Science Quarterly* 114:1 (1999), 53–78.

62. Ian Lustick, *Trapped in the War on Terror* (Philadelphia: University of Pennsylvania Press, 2006); John Mueller, *Overblown: How Politicians and the Terrorism Industry Inflate National Security Threats, and Why We Believe Them* (New York: Simon & Schuster, 2009); Jason Ralph, *America's War on Terror: The State of the 9/11 Exception from Bush to Obama* (Oxford: Oxford University Press, 2013).

Chapter 4

1. Christopher S. Chivvis, *Toppling Qaddafi: Libya and the Limits of Liberal Intervention* (New York: Cambridge University Press, 2014); Michael P. Kreuzer, *Drones and the Future of Air Warfare: The Evolution of Remotely Piloted Aircraft* (New York: Routledge, 2016).

2. Brian Glyn Williams, *Afghanistan Declassified: A Guide to America's Longest War* (Philadelphia: University of Pennsylvania Press, 2012).

3. Michael C. Horowitz, Sarah E. Kreps, and Matthew Fuhrmann, "Separating

Fact from Fiction in the Debate over Drone Proliferation," *International Security* 41:2 (2016).

4. Sharon Weinberger, "The Ultra-Lethal Drones of the Future," *New York Post*, May 17, 2014, http://nypost.com/2014/05/17/evolution-of-the-drone/; P. W. Singer, "The Future of War Will Be Robotic," *CNN*, February 23, 2015, http://edition.cnn.com/2015/02/23/opinion/singer-future-of-war-robotic/; Kris Osborn, "Navy Secretary Says Future Navy Fighter Planes Will Be Unmanned," *Military.com*, April 16, 2015, http://www.military.com/daily-news/2015/04/16/navy-secretary-says-future-navy-fighter-planes-will-be-unmanned.html

5. For examples see Christian Enemark, *Armed Drones and the Ethics of War: Military Virtue in a Post-Heroic Age* (New York: Routledge, 2013); John Kaag and Sarah Kreps, *Drone Warfare* (Malden, MA: Polity Press, 2014); Grégorie Chamayou, *A Theory of the Drone* (New York: The New Press, 2015).

6. Michael Walzer, *Just and Unjust Wars: A Moral Argument with Historical Illustrations* (New York: Basic Books, 1977); Brian Orend, *The Morality of War* (Orchard Park, NY: Broadview Press, 2006); Steven Lee, *Ethics and War: An Introduction* (New York: Cambridge University Press, 2012).

7. Daniel Brunstetter and Megan Braun, "The Implications of Drones on the Just War Tradition," *Ethics & International Affairs* 25:3 (2011), 337–58.

8. Zack Beauchamp and Julian Savulescu, "Robot Guardians: Tele-operated Combat Vehicles in Humanitarian Military Intervention," in *Killing by Remote Control: The Ethics of an Unmanned Military*, ed. Bradley Jay Strawser (New York: Oxford University Press, 2013), 106.

9. Bruce Jentleson, "The Pretty Prudent Public: Post Post-Vietnam American Opinion on the Use of Military Force," *International Studies Quarterly* 36 (1992), 49–73.

10. Walzer, *Just and Unjust Wars*; Orend, *The Morality of War*; Lee, *Ethics and War*.

11. Deen K. Chatterjee, ed., *The Ethics of Preventive War* (Cambridge: Cambridge University Press, 2013).

12. Daniel W. Drezner, "The Realist Tradition in American Public Opinion," *Perspectives on Politics* 1 (2008), 51–70.

13. Bruce W. Jentleson and Rebecca L. Britton, "Still Pretty Prudent: Post-Cold War American Public Opinion on the Use of Military Force," *Journal of Conflict Resolution* 42:4 (1998), 395–417.

14. Christopher Gelpi, Peter D. Feaver, Jason Reifler, *Paying the Human Costs of War: American Public Opinion and Casualties in Military Conflicts* (Princeton: Princeton University Press, 2009), 65.

15. Gelpi, Feaver, and Reifler, *Paying the Human Costs of War*, 107.

16. A. Cooper Drury, L. Marvin Overbury, Adrian Ang, Yitan Li, "'Pretty Prudent' or Rhetorically Responsive? The American Public's Support for Military Action," *Political Research Quarterly* 63:1 (2010), 92.

17. For example, when Buchanan and Keohane propose a regulatory regime that would apply to drone operations, they are almost exclusively concerned with the challenges related to the targeted killing of suspected terrorists. See Allen Buchanan and Robert O. Keohane, "Toward a Drone Accountability Regime," *Ethics & International Affairs* 29:1 (2015).

18. Suzy Killmister, "Remote Weaponry: The Ethical Implications," *Journal of Applied Philosophy* 25:2 (2008), 121–33; Jai C. Galliott, "Uninhabited Aerial Vehicles and the Asymmetry Objection: A Response to Strawser," *Journal of Military Ethics* 11:1 (2012), 58–66.

19. John Mueller, *War, Presidents, and Public Opinion* (New York: Wiley, 1973).

20. Hugh Smith, "What Costs Will Democracies Bear? A Review of Popular Theories of Casualty Aversion," *Armed Forces & Society* 31:4 (2005), 493.

21. Hugh Smith, "What Costs Will Democracies Bear?" 494.

22. Gelpi, Feaver, and Reifler, *Paying the Human Costs of War*, 154.

23. Respondents were recruited via Mechanical Turk and completed the experiment in January 2015. See the chapter appendix for demographic details.

24. Feaver, Gelpi, and Reifler, *Paying the Human Costs of War*.

25. Shibley Telhami, "American Attitudes toward the Middle East," *The Center for Middle East Policy at the Brookings Institution*, July 11, 2016. https://www.brookings.edu/wp-content/uploads/2016/07/FINAL-KEY-FINDINGS-3.pdf

26. Jeff Mason, "A Third of Republicans Support Iran Nuclear Deal," *Reuters*, April 8, 2015, https://www.reuters.com/article/us-iran-nuclear-usa-poll/a-third-of-republicans-support-iran-nuclear-deal-reuters-ipsos-poll-idUSKBN0MZ0AG20150408 (accessed October 8, 2017).

27. Matthew Evangelista, *Law, Ethics, and the War on Terror* (Malden, MA: Polity, 2008).

28. For more discussion of the tone of research on casualty aversion and its normative implications, see Smith, "What Costs Will Democracies Bear?"

29. Medea Benjamin, *Drone Warfare: Killing by Remote Control* (New York: OR Books, 2012); Enemark, *Armed Drones and the Ethics of War*.

30. Disagreement over whether the 2003 invasion of Iraq was justified is evidence of this. See Jean Bethke Elshtain, *Just War Against Terror* (New York: Basic Books, 2003); Neta C. Crawford, "Just War Theory and the U.S. Counterterror War," *Perspectives on Politics* 1:1 (2003), 5–25.

31. Bruce Bueno de Mesquita and Randolph Siverson, "War and the Survival of Political Leaders: A Comparative Study of Regime Types and Political Accountability," *American Political Science Review* 89:4 (1995), 841–55; Hein Goemans, "Fighting for Survival: The Fate of Leaders and the Duration of War," *Journal of Conflict Resolution* 44:5 (2000), 555–79; David L. Rousseau, A. Trevor Thrall, Marcus Schulzke, and Steve S. Sin, "Democratic Leaders and War: Simultaneously Managing External Conflicts and Domestic Politics," *Australian Journal of International Affairs* 66:3 (2012), 349–64.

32. D. Sunshine Hillygus and Todd G. Shields, "Moral Issues and Voter Decision Making in the 2004 Presidential Election," *PS: Political Science & Politics* 38 (2005), 201–9; Yotam Margalit, "Explaining Social Policy Preferences: Evidence from the Great Recession," *American Political Science Review* 107:1 (2013), 80–103; Jacob S. Hacker, Philipp Rehm, and Mark Schlesinger, "The Insecure American: Economic Experiences, Financial Worries, and Policy Attitudes," *Perspectives on Politics* 11:1 (2013), 23–49.

33. Frank Sauer and Niklas Schörnig, "Killer Drones: The 'Silver Bullet' of Democratic Warfare?" *Security Dialogue* 43:4 (2012).

34. Adam J. Berinsky, *In Time of War: Understanding American Public Opinion from World War II to Iraq* (Chicago: University of Chicago Press, 2009), 69.

35. Michael J. Boyle, "The Costs and Consequences of Drone Warfare," *International Affairs* 89:1 (2013), 1–29; David Hastings Dunn, "Drones: Disembodied Aerial Warfare and the Unarticulated Threat," *International Affairs* 89:5 (2013), 1237–46; Enemark, *Armed Drones and the Ethics of War.*

Chapter 5

1. Christopher Gelpi, Peter D. Feaver, and Jason Reifler, "Success Matters: Casualty Sensitivity and the War in Iraq," *International Security* 30:3 (2006), 7–46; Christopher Gelpi, Peter D. Feaver, Jason Reifler, *Paying the Human Costs of War: American Public Opinion & Casualties in Military Conflicts* (Princeton: Princeton University Press, 2009).

2. John Kaag and Sarah Kreps, *Drone Warfare* (Malden, MA: Polity Press, 2014), 108.

3. Gelpi, Feaver, and Reifler, *Paying the Human Costs of War*, 123.

4. Peter D. Feaver and Christopher Gelpi, *Choosing Your Battles: American Civil-Military Relations and the Use of Force* (Princeton: Princeton University Press, 2004), 234.

5. Feaver and Gelpi, *Choosing Your Battles*, 200.

6. Scott Sigmund Gartner, "The Multiple Effects of Casualties on Public Support for War: An Experimental Approach," *American Political Science Review* 102:1 (2008), 95–106.

7. Adam J. Berinsky, *In Time of War: Understanding American Public Opinion from World War II to Iraq* (Chicago: University of Chicago Press, 2009), 103.

8. See also Adam J. Berinsky, "Assuming the Costs of War: Events, Elites, and American Public Support for Military Conflict," *Journal of Politics* 69:4 (2007), 975–97.

9. Adam J. Berinsky, James N. Druckman, "Public Opinion Research and Support for the Iraq War," *Public Opinion Quarterly* 71:1 (2007), 126–41.

10. Christopher Gelpi, "Performing on Cue? The Formation of Public Opinion Toward War," *Journal of Conflict Resolution* 54:1 (2010), 88–116.

11. Ronald C. Arkin, *Governing Lethal Behavior in Autonomous Robots* (Boca Raton, FL: Taylor & Francis Group, 2009); Bradley Jay Strawser, "Moral Predators: The Duty to Employ Uninhabited Aerial Vehicles," *Journal of Military Ethics* 9:4 (2010), 342–68; and Marcus Schulzke, "The Morality of Remote Warfare: Against the Asymmetry Objection to Remote Weaponry," *Political Studies* (2014).

12. Arkin, *Governing Lethal Behavior in Autonomous Robots*; Schulzke, "The Morality of Remote Warfare"; Marcus Schulzke, "Robots as Weapons in Just Wars," *Philosophy and Technology* 24:3 (2011), 293–306; Anton Petrenko, "Between Berserksgang and the Autonomous Weapons Systems," *Public Affairs Quarterly* 26:2 (2012), 81–102.

13. Daniel Byman, "Why Drones Work: The Case for Washington's Weapon of Choice," *Foreign Affairs* (July/August 2013).

14. J. Marshall Beier, "Discriminating Tastes: 'Smart' Bombs, Non-Combatants, and Notions of Legitimacy in Warfare," *Security Dialogue* 34 (2003), 411–25.

15. Scott Shane, *Objective Troy: A Terrorist, a President, and the Rise of the Drone* (New York: Tim Duggan Books, 2015).

16. Spencer Ackerman, "Obama Claims US Drones Strikes Have Killed Up to 116 Civilians," *The Guardian*, July 1, 2016, available at https://www.theguardian.com/us-news/2016/jul/01/obama-drones-strikes-civilian-deaths

17. David Leigh, "Iraq War Logs Reveal 15,000 Previously Unlisted Civilian Deaths—Leaked Pentagon Files Contain Records of More than 100,000 Fatalities Including 66,000 Civilians," *The Guardian*, October 22, 2010.

18. For example see https://www.iraqbodycount.org/

19. Robert Cassidy, *Counterinsurgency and the Global War on Terror: Military Culture and Irregular War* (Stanford, CA: Stanford University Press, 2008); David Kilcullen, T*he Accidental Guerrilla: Fighting Small Wars in the Midst of a Big One* (New York: Oxford University Press, 2009).

20. FM 3-24, *Counterinsurgency, FM 3-24/MCWP 3-33.5* (Washington, DC: Headquarters Department of the Army, 2006).

21. Michael C. Horowitz, Sarah E. Kreps, and Matthew Fuhrmann, "Separating Fact from Fiction in the Debate over Drone Proliferation," *International Security* 41:2 (2016).

22. As an example of these projects see https://www.defense.gov/News/Contracts/Contract-View/Article/1198370/ (accessed August 10, 2017).

23. Byman, "Why Drones Work"; Leon Panetta and Jim Newton, *Worthy Fights: A Memoir of Leadership in War and Peace* (New York: Penguin, 2014).

24. "Obama's Speech on Drone Policy," *New York Times*, May 23, 2013, http://www.nytimes.com/2013/05/24/us/politics/transcript-of-obamas-speech-on-drone-policy.html

25. David Kilcullen, Andrew McDonald Exum, "Death from Above, Outrage Down Below," *New York Times*, May 16, 2009, http://www.nytimes.com/2009/05/17/opinion/17exum.html?pagewanted=all&_r=0

26. Derek Gregory, "From a View to a Kill: Drones and Late Modern War," *Theory, Culture, & Society* 28:7–8 (2011), 188–215; Jeffrey Sluka, "Death from Above: UAVs and Losing Hearts and Minds," *Military Review* (May–June 2011), 70–76.

27. Michael J. Boyle, "The Costs and Consequences of Drone Warfare," *International Affairs* 89:1 (2013), 1–29, 14.

28. Suzy Killmister, "Remote Weaponry: The Ethical Implications," *Journal of Applied Philosophy* 25:2 (2008), 121–33; Christian Enemark, *Armed Drones and the Ethics of War: Military Virtue in a Post-Heroic Age* (New York: Routledge, 2013).

29. Ariel Rubinstein and Menahem E. Yaari, "Repeated Insurance Contracts and Moral Hazard," *Journal of Economic Theory* 30 (1983), 74–97, 74.

30. Gary H. Stern and Ron J. Feldman, *Too Big to Fail: The Hazards of Bank Bailouts* (Washington, DC: Brookings Institution, 2004).

31. Carol A. Heimer, *Reactive Risk and Rational Action: Managing Moral Hazard in Insurance Contracts* (Berkeley: University of California Press, 1985).

32. Sam Peltzman, "The Effects of Automobile Safety Regulation," *Journal of Political Economy* 83:4 (1975), 677–726.

33. Greg Ip, *Foolproof: Why Safety Can Be Dangerous and How Danger Makes Us Safe* (New York: Little, Brown, 2015).

34. John Kaag and Sarah Kreps, *Drone Warfare* (Malden, MA: Polity Press, 2014), 109.

35. Paul W. Kahn, "The Paradox of Riskless Warfare," *Philosophy and Public Policy Quarterly* 22:3 (2002), 2–8.

36. Grégorie Chamayou, *A Theory of the Drone* (New York: The New Press, 2015).

37. Martin Shaw, "Risk-Transfer Militarism, Small Massacres and the Historic Legitimacy of War," *International Relations* 16:3 (2002), 343–60; Martin Shaw, *The New Western Way of War: Risk-Transfer War and Its Crisis in Iraq* (Malden, MA: Polity Press, 2005).

38. Jean Baudrillard, *The Gulf War Did Not Take Place* (Bloomington: Indiana University Press, 1995).

39. Michael Ignatieff, *Virtual War: Kosovo and Beyond* (New York: Picador, 2001).

40. James Der Derian, "Virtuous War/Virtual Theory," *International Affairs* 76:4 (2000), 771–88; James Der Derian, *Virtuous War: Mapping the Military-Industrial-Media-Entertainment Network* (New York: Routledge, 2009).

41. Christopher Coker, *Waging War Without Warriors? The Changing Culture of Military Conflict* (Boulder: Lynne Rienner, 2002).

42. Colin McInnes, *Spectator Sport War: The West and Contemporary Conflict* (Boulder: Lynne Rienner, 2002).

43. Richard N. Haas, *War of Necessity, War of Choice: A Memoir of Two Iraq Wars* (New York: Simon & Schuster, 2009), 10.

44. At the same time, the public might recognize that decisions about which type of force to employ in a particular situation is a strategic one. Political and military leaders might choose, for example, to deploy drones against targets that are more difficult or costly to attack with manned aircraft or ground troops. In the case of the data summarized in figure 5.1, this recognition of strategic interaction between target characteristics and attack type is less likely to be driving estimates of success, as the respondents in this experiment were all presented with identical information about the target. Nonetheless, how such characteristics influence estimates of success is a topic that warrants further research.

45. The texts of the treatments, survey instrument, and demographic information about the respondents can be found in the chapter appendix.

46. Aaron M. Hoffman, Christopher R. Agnew, Laura E. VanderDrift, and Robert Kulzick, "Norms, Diplomatics Alternatives, and the Social Psychology of War Support," *Journal of Conflict Resolution* 59:1 (2013), 3–28, quotations from page 5. See also Benjamin I. Page and Marshal M. Bouton, *The Foreign Policy Disconnect: What Americans Want from Our Leaders But Don't Get* (Chicago: University of Chicago Press, 2006).

47. $\chi^2 = 10.92$, degrees of freedom = 15, $p = .76$.

48. Frances V. Harbour, "Reasonable Probability of Success as a Moral Criterion in the Just War Tradition," *Journal of Military Ethics* 10:3 (2011), 230–41.

49. Alexander B. Downes, *Targeting Civilians in War* (Ithaca: Cornell University Press, 2008), 3.

50. Chamayou, *A Theory of the Drone*, 77.

Chapter 6

1. John Mueller, "Public Opinion as a Constraint on U.S. Foreign Policy: Assessing the Perceived Value of American and Foreign Lives" (manuscript, 2000); John Tirman, *The Deaths of Others: The Fate of Civilians in America's Wars* (New York: Oxford University Press, 2011).

2. Richard Eichenberg, "Victory Has Many Friends: U.S. Public Opinion and the Use of Force," *International Security* 30 (2005), 140–77.

3. Robert Johns and Graeme A. M. Davies, "Civilian Casualties and Public Support for Military Action: Experimental Evidence," *Journal of Conflict Resolution* (forthcoming).

4. Benjamin Valentino, "Moral Character or Character of War? American Public Opinion on the Targeting of Civilians in War," *Daedalus* 145:4 (2016), 127–38.

5. Harvey Sapolsky and Jeremy Shapiro, "Casualties, Technology, and America's Future Wars," *Parameters* (1996), 119–27.

6. Daniel Kahneman and Amos Tversky, "The Simulation Heuristic," in *Judgement Under Uncertainty: Heuristics and Biases*, ed. Daniel Kahneman, Amos Tversky, and Paul Slovic (New York: Cambridge University Press, 1982), 201–8.

7. Ruth M. J. Byrne, *The Rational Imagination: How People Create Alternatives to Reality* (Cambridge, MA: MIT Press, 2005); Neil J. Roese, "Counterfactual Thinking," *Psychological Bulletin* 121:1 (1997), 133–48.

8. Dale T. Miller and Brian R. Taylor, "Counterfactual Thought, Regret, and Superstition: How to Avoid Kicking Yourself," in *What Might Have Been: The Social Psychology of Counterfactual Thinking*, ed. Neil J. Roese and James M. Olson (Hillsdale, NJ: Lawrence Erlbaum Associates, 1995), 305–11.

9. Mark D. Alicke, Justin Buckingham, Ethan Zell, and Teresa Davis, "Culpable Control and Counterfactual Reasoning in the Psychology of Blame," *Personality & Social Psychology Bulletin* 34:10 (2008), 1371–81; Dale T. Miller and Charles McFarland, "Counterfactual Thinking and Victim Compensation: A Test of Norm Theory," *Personality and Social Psychology Bulletin* 12 (1986), 513–19.

10. Roese, "Counterfactual Thinking."

11. V. H. Medvec, S. F. Madey, and T. Gilovich, "When Less Is More: Counterfactual Thinking and Satisfaction Among Olympic Medalists," *Journal of Personality and Social Psychology* 69:4 (1995), 603–10.

12. Daniel Kahneman and Dale T. Miller, "Norm Theory: Comparing Reality to Its Alternatives," *Psychological Review* 93:2 (1986), 136–53.

13. Christopher Gelpi, Peter Feaver, and Jason Reifler, *Paying the Human Costs of War: American Public Opinion and Casualties in Military Conflicts* (Princeton: Princeton University Press, 2009).

14. Chris Woods, Jack Serle, and Alice Ross, "Emerging from the Shadows: US Covert Drone Strikes in 2012," *The Bureau of Investigative Journalism*, January 3, 2013, http://www.thebureauinvestigates.com/2013/01/03/emerging-from-the-shadows-us-covert-drone-strikes-in-2012-2/

15. We deliberately did not ask in the first round of this experiment questions about how concerned the respondent was about civilian casualties, since this might have primed attention to such casualties in subsequent information and questions. Instead,

we measure concern about civilian harm indirectly by comparing the overall level of support for the attack where the only factor that varies is likelihood of civilian casualties.

16. The United States military has regularly made such "condolence payments" to families of civilians killed as a consequence of American military operations. During confirmation hearings for the position of director of the Central Intelligence Agency in 2013, John Brennan stated that condolence payments are offered to families of victims of drone strikes: "In those rare instances in which civilians have been killed [in drone strikes], after-action reviews have been conducted to identify corrective actions and to minimize the risk of innocents being killed or injured in the future. Where possible, we also work with local governments to gather facts and, if appropriate, provide condolence payments to families of those killed." See "Questions for the Record, Mr. John Brennan, Questions from the Chairman," Senate Select Committee on Intelligence, n.d., available at http://www.intelligence.senate.gov/130207/posthearing.pdf; and Cora Currier, "Drones In Yemen: Does the U.S. Pay Families When Strikes Kill Innocent Yemenis?," *Huffington Post*, August 12, 2013.

17. See Miller and McFarland, "Counterfactual Thinking and Victim Compensation." This experiment uses a between-respondents design; see Eric van Dijk and Marcel Zeelenberg, "On the Psychology of 'if Only': Regret and the Comparison Between Factual and Counterfactual Outcomes," *Organizational Behavior and Human Decision Processes* 97 (2005), 152–60.

18. Results reported here are unchanged when the data is analyzed with logistic regression models that include control variables; see the chapter appendix for details.

19. For more on US compensatory payments in recent wars, see United States Government Accountability Office, *Military Operations: The Department of Defense's Use of Solatia and Condolence Payments in Iraq and Afghanistan, GAO-07-699* (2007); David Zucchino, "U.S. 'Condolence Payments' Translate Iraqis' Losses to Cash," *The Seattle Times*, March 12, 2005; Amanda Carroll and Marcus Schulzke, "Compensating Civilians during War: A Place for Individuals in International Law," *Democracy and Security* 9:4 (2013), 398–421; Marcus Schulzke and Amanda Cortney Carroll, "Compensation and the Right to Life: Corrective Justice for the Civilian Victims of War," *Journal of International Relations and Development* (2015).

20. See the replication materials for the details of these statistical tests.

21. Jacquelyn Schneider and Julia Macdonald, *U.S. Public Support for Drone Strikes: When Do Americans Prefer Unmanned Over Manned Platforms?* (Washington, DC: Center for a New American Security, 2015).

22. Eric Larson and Bogdan Savych, *Misfortunes of War: Press and Public Reactions to Civilian Deaths in Wartime* (Santa Monica, CA: RAND, 2006). This differs from cases in which political and military leaders in the United States and other countries have deliberately targeted civilians or ignored how their tactics place civilians in danger; see Bruce Cronin, "Reckless Endangerment Warfare: Civilian Casualties and the Collateral Damage Exception in International Humanitarian Law," *Journal of Peace Research* 50:2 (2013), 175–87; Alexander B. Downes, *Targeting Civilians in War* (Ithaca: Cornell University Press, 2008); Alexander B. Downes, "Desperate Times, Desperate Measures: The Causes of Civilian Victimization in War," *International Security* 30:4 (2006), 152–95; Colin H. Kahl, "In the Crossfire or the Crosshairs? Norms,

Civilian Casualties, and U.S. Conduct in Iraq," *International Security* 32:1 (2007), 7–46; Nick Turse, *Kill Anything That Moves: The Real American War in Vietnam* (New York: Metropolitan Books, 2013). Kahl, in "In the Crossfire," argues that the United States military took steps after the Vietnam War to increase its compliance with international humanitarian war, and that these steps likely reduced but did not eliminate unnecessary harm to civilians during the Iraq War that began in 2003.

23. Cora Currier, "How Does the U.S. Mark Unidentified Men in Pakistan and Yemen as Drone Targets?," *ProPublica*, March 1, 2013, http://reason.com/poll/2013/03/01/february-2013-national-survey; Kevin Heller, "'One Hell of a Killing Machine': Signature Strikes and International Law," *Journal of International Criminal Justice* 89 (2013).

24. This norm is typically called "discrimination" in just war theory and "distinction" in international law.

25. Michael Waltzer, *Just and Unjust Wars: A Moral Argument with Historical Illustrations*, 4th ed. (New York: Basic Books, 2006).

26. Treaty texts available at the International Committee of the Red Cross' international humanitarian law database, available at http://www.icrc.org/ihl

27. Quotations from the 1977 First Additional Protocol and Y. Sandoz, C. Swinarski, and B. Zimmerman, eds., *Commentary on the Additional Protocols of 8 June 1977 to the Geneva Conventions of 12 August 1949* (The Hague: Martinus Nijhoff, 1987), 2195.

28. Michael X. Delli Carpini, *What Americans Know About Politics and Why It Matters* (New Haven: Yale University Press, 1996); John Zaller, *The Nature and Origins of Mass Opinion* (New York: Cambridge University Press, 1992); Adam J. Berinsky, "Assuming the Costs of War: Events, Elites, and American Public Support for Military Conflict," *Journal of Politics* 69:4 (2007), 975–97.

29. Arthur Lupia and Matthew McCubbins, *The Democratic Dilemma: Can Citizens Learn What They Need to Know?* (New York: Cambridge University Press, 1998).

30. Tomz, "Reputation and the Effect of International Law on Preferences and Beliefs" (manuscript, 2008).

31. Geoffrey P. R. Wallace, "International Law and Support for Torture: An Experimental Study," *International Organization* 67:1 (2012), 105–40.

32. Rose McDermott, "Prospect Theory in International Relations: The Iranian Hostage Rescue Mission," *Political Psychology* 13:2 (1992), 237–63.

Chapter 7

1. Erik Voeten and Paul R. Brewer, "Public Opinion, the War in Iraq, and Presidential Accountability," *Journal of Conflict Resolution* 50:6 (2006), 809–30; Michael T. Heaney and Fabio Rojas, "Partisans, Nonpartisans, and the Antiwar Movement in the United States," *American Politics Research* 35:4 (2007), 431–64; Douglas Kriner and Francis Shen, "Responding to War on Capitol Hill: Battlefield Casualties, Congressional Response, and Public Support for the War in Iraq," *American Journal of Political Science* 58:1 (2014), 157–74; Benny Geys, "Wars, Presidents, and Popularity: The Political Cost(s) of War Re-Examined," *Public Opinion Quarterly* 74:2 (2010), 357–74.

2. Michael O'Hanlon, "Can High Technology Bring U.S. Troops Home?," *Foreign Policy* 113 (1998–1999), 72–86.

3. David A. Koplow, *Death by Moderation: The U.S. Military's Quest for Useable Weapons* (New York: Cambridge University Press, 2010).

4. For more on precision weapons and other forms of selective violence, see David A. Koplow, *Non-Lethal Weapons: The Law and Policy of Revolutionary Technologies for the Military and Law Enforcement* (Cambridge: Cambridge University Press, 2006); Koplow, *Death by Moderation.*

5. Important exceptions include Sarah E. Kreps, "Flying under the Radar: A Study of Public Attitudes Towards Unmanned Aerial Vehicles," *Research & Politics* 1:1 (2014), 1–7; and Michael C. Horowitz, "Public Opinion and the Politics of the Killer Robot Debate," *Research & Politics* 3:1 (2014), 1–8.

6. John H. Aldrich, John L. Sullivan, and Eugene Borgida, "Foreign Affairs and Issue Voting: Do Presidential Candidates 'Waltz Before A Blind Audience?'" *American Political Science Review* 83:1 (1989), 123–41; Matthew A. Baum, "How Public Opinion Constrains the Use of Force: The Case of Operation Restore Hope," *Presidential Studies Quarterly* 34:2 (2004), 187–26, Jon A. Krosnick, and Donald R. Kinder, "Altering the Foundations of Support for the President Through Priming," *American Political Science Review* 84:2 (1989), 497–512; and Richard Sobel, *The Impact of Public Opinion on U.S. Foreign Policy Since Vietnam: Constraining the Colossus* (New York: Oxford University Press, 2001).

7. David L. Rousseau, Trevor Thrall, Marcus Schulzke, and Steve S. Sin, "Democratic Leaders and War: Simultaneously Managing External Conflicts and Domestic Politics," *Australian Journal of International Affairs* 66:3 (2012), 349–64.

8. Christopher Hood, *The Blame Game: Spin, Bureaucracy, and Self-Preservation in Government* (Princeton: Princeton University Press, 2010).

9. Dan Eggen, "Citing History, Bush Suggests His Policies Will One Day Be Vindicated," *Washington Post*, June 9, 2008, http://www.washingtonpost.com/wp-dyn/content/article/2008/06/08/AR2008060802255.html

10. Alan I. Abramowitz and Kyle L. Saunders, "Is Polarization a Myth?" *Journal of Politics* 70:2 (2008), 542–55.

11. Bruce Jentleson, "The Pretty Prudent Public: Post Post-Vietnam American Opinion on the Use of Military Force," *International Studies Quarterly* 36 (1992), 49–73; Bruce W. Jentleson and Rebecca L. Britton, "Still Pretty Prudent: Post-Cold War American Public Opinion on the Use of Military Force," *Journal of Conflict Resolution* 42:4 (1998), 395–417.

12. Christopher Gelpi, Peter Feaver, and Jason Reifler, "Success Matters: Casualty Sensitivity and the War in Iraq," *International Security* 30:3 (2006), 7–46; Christopher Gelpi, Peter Feaver, and Jason Reifler, *Paying the Human Costs of War: American Public Opinion and Casualties in Military Conflicts* (Princeton: Princeton University Press, 2009).

13. Allen Buchanan and Robert O. Keohane, "Toward a Drone Accountability Regime," *Ethics & International Affairs* 29:1 (2015), 17.

14. Michael J. Boyle, "The Costs and Consequences of Drone Warfare," *International Affairs* 89:1 (2013), 1–29.

15. Zack Beauchamp and Julian Savulescu, "Robot Guardians: Teleoperated Com-

bat Vehicles in Humanitarian Military Operations," in *Killing by Remote Control: The Ethics of an Unmanned Military*, ed. Bradley Jay Strawser (New York: Oxford University Press, 2013).

16. For more detail on al Awlaki and the decision to kill him, see Scott Shane, *Objective Troy: A Terrorist, a President, and the Rise of the Drone* (New York: Tim Duggan Books, 2015).

17. Glenn Greenwald, "The Due-Process-Free Assassination of U.S. Citizens Is Now Reality," *Salon*, September 30, 2011, http://www.salon.com/2011/09/30/awlaki_6/

18. Buchanan and Keohane, "Toward a Drone Accountability Regime."

19. Chris Woods, *Sudden Justice: The True Costs of America's Secret Drone War* (London: C. Hurst, 2014), 260–65.

20. Woods, *Sudden Justice*, 205.

21. Shane, *Objective Troy*; Woods, *Sudden Justice*.

22. See https://www.thebureauinvestigates.com/category/projects/drones/drones-graphs/

23. Jack Serle, "Obama Drone Casualty Numbers a Fraction of Those Recorded by the Bureau," *Bureau of Investigative Journalism*, July 1, 2016, https://www.thebureauinvestigates.com/2016/07/01/obama-drone-casualty-numbers-fraction-recorded-bureau/

24. Amanda Carroll and Marcus Schulzke, "Compensating Civilians during War: A Place for Individuals in International Law," *Democracy and Security* 9:4 (2013), 398–421; Marcus Schulzke, *Protecting the Victims of War: Just War Theory and the Duty to Assist Civilian Casualties.* (Cambridge: Cambridge University Press, 2007).

25. "Privacy Impact Assessment for the Aircraft Systems," Department of Homeland Security, DHS/CBP/PIA-018, September 9, 2013.

26. Concept of Operations for CBP's Predator B Unmanned Aircraft System: Fiscal Year 2010 Report to Congress, U.S. Customs and Border Protection, June 29, 2010.

27. For more on the confusion of war-fighting and law-enforcement paradigms, see Paul W. Kahn, "The Paradox of Riskless Warfare," *Philosophy and Public Policy Quarterly* 22:3 (2002), 2–8.

28. Ronald C. Arkin, *Governing Lethal Behavior in Autonomous Robots* (Boca Raton, FL: Taylor & Francis, 2009); Marcus Schulzke, "Autonomous Weapons and Distributed Responsibility," *Philosophy and Technology* 26:2 (2013), 203–19; Aaron Johnson and Sidney Axinn, "The Morality of Autonomous Robots," *Journal of Military Ethics* 12:2 (2013), 129–41; James Igoe Walsh, "Political Accountability and Autonomous Weapons," *Research and Politics* 2:4 (2016), 1–6.

29. Horowitz, "Public Opinion and the Politics of the Killer Robots Debate"; Michael C. Horowitz, "The Ethics and Morality of Robotic Warfare: Assessing the Debate Over Autonomous Weapons," *Daedalus* 145:4 (2016), 25–36; James Igoe Walsh, "Political Accountability and Autonomous Weapons," *Research & Politics* 2:4 (2015), 1–6.

30. Simon Rogers, "Drones by Country: Who Has All the UAVs?" *The Guardian*, August 3, 2012, http://www.theguardian.com/news/datablog/2012/aug/03/drone-stocks-by-country

31. Paul Fahlstrom and Thomas Gleason, *Introduction to UAV Systems* (New York: Wiley, 2012), 5.

32. Woods, *Sudden Justice*, 52.

33. Scott Wilson, "In Gaza, Lives Shaped by Drones," *Washington Post*, December 3, 2011, https://www.washingtonpost.com/world/national-security/in-gaza-lives-shaped-by-drones/2011/11/30/gIQAjaP6OO_story.html

34. Ewen MacAskill, "Drone Killing of British Citizens in Syria Marks Major Departure for UK," *The Guardian*, September 7, 2015, https://www.theguardian.com/world/2015/sep/07/drone-british-citizens-syria-uk-david-cameron

35. Andrew Rettman, "Seven EU States Create Military Drone 'Club,'" November 20, 2013, https://euobserver.com/defence/122167

36. Frank Sauer and Niklas Schörnig, "Killer Drones: The 'Silver Bullet' of Democratic Warfare?" *Security Dialogue* 43:4 (2012), 363–80, 369.

37. Michael C. Horowitz and Matthew Fuhrmann, "Droning On: The Proliferation of Unmanned Aerial Vehicles and Military Innovation," *International Organization* (forthcoming).

38. Fuhrmann and Horowitz, "Droning On"; Andrea Gilli and Mauro Gilli, "The Diffusion of Drone Warfare? Industrial, Organizational, and Infrastructural Constraints," *Security Studies* 25:1 (2016), 50–84; Michael C. Horowitz, Sarah E. Kreps, and Matthew Fuhrmann, "Separating Fact from Fiction in the Debate over Drone Proliferation," *International Security* 41:2 (2016), 7–42.

Appendix to Chapter 6

1. The survey instrument includes items about the respondent's political knowledge. We sought to use these items to create an index of political knowledge, but Cronbach's alpha indicated that this index had low internal consistency.

2. Eric Larson and Bogdan Savych, *Misfortunes of War: Press and Public Reactions to Civilian Deaths in Wartime* (Santa Monica, CA: RAND, 2006).

Index